Lecture Notes in Civil Engineering

Volume 161

Lecture Notes in Civil Engineering (LNCE) publishes the latest developments in Civil Engineering - quickly, informally and in top quality. Though original research reported in proceedings and post-proceedings represents the core of LNCE, edited volumes of exceptionally high quality and interest may also be considered for publication. Volumes published in LNCE embrace all aspects and subfields of, as well as new challenges in, Civil Engineering. Topics in the series include:

- Construction and Structural Mechanics
- Building Materials
- Concrete, Steel and Timber Structures
- Geotechnical Engineering
- Earthquake Engineering
- Coastal Engineering
- Ocean and Offshore Engineering; Ships and Floating Structures
- Hydraulics, Hydrology and Water Resources Engineering
- Environmental Engineering and Sustainability
- Structural Health and Monitoring
- Surveying and Geographical Information Systems
- Indoor Environments
- Transportation and Traffic
- Risk Analysis
- Safety and Security

To submit a proposal or request further information, please contact the appropriate Springer Editor:

- Pierpaolo Riva at pierpaolo.riva@springer.com (Europe and Americas);
- Swati Meherishi at swati.meherishi@springer.com (Asia - except China, and Australia, New Zealand);
- Wayne Hu at wayne.hu@springer.com (China).

All books in the series now indexed by Scopus and EI Compendex database!

More information about this series at http://www.springer.com/series/15087

Lin Yola · Utaberta Nangkula ·
Olutobi Gbenga Ayegbusi ·
Mokhtar Awang
Editors

Sustainable Architecture and Building Environment

Proceedings of ICSDEMS 2020

Springer

Editors
Lin Yola
Universitas Kristen Indonesia
Jakarta, Indonesia

Olutobi Gbenga Ayegbusi
Kolej Universiti Linton
Mantin, Negeri Sembilan, Malaysia

Utaberta Nangkula
Faculty of Civil Engineering and Built Environment
Universiti Tun Hussein Onn Malaysia
Parit Raja, Johor, Malaysia

Mokhtar Awang
Department of Mechanical Engineering
Universiti Teknologi Petronas
Seri Iskander, Perak, Malaysia

ISSN 2366-2557 ISSN 2366-2565 (electronic)
Lecture Notes in Civil Engineering
ISBN 978-981-16-2328-8 ISBN 978-981-16-2329-5 (eBook)
https://doi.org/10.1007/978-981-16-2329-5

This Springer imprint is published by the registered company Springer Nature Singapore Pte Ltd.
The registered company address is: 152 Beach Road, #21-01/04 Gateway East, Singapore 189721, Singapore

Preface

This book presents a compilation of research works of ICSDEMS 2020.

School of Strategic and Global Studies, Universitas Indonesia, is proud to host and organize the second edition of the annual International Conference on Sustainable Design, Engineering, Management, and Sciences (ICSDEMS 2020) that held during December 8–9, 2020, through virtual presentations.

Conference aims to provide an international platform for researchers, academicians, students as well as industrial professionals from all over the globe to present, share, and exchange their scientific research outcomes and developmental activities about all parts of green built environment. This conference will include various specializations related to planning, building, engineering, information technology, project management, and social science that significantly support the development of sustainable built environment. This conference provides and delegates a platform for knowledge sharing and discussion, establishment of research networking, and enabling a global partner for future research collaboration.

The editor(s) of the proceeding would like to express the utmost gratitude and thanks to all reviewers in the technical team for making this volume a success.

Jakarta, Indonesia — Lin Yola
Parit Raja, Malaysia — Utaberta Nangkula
Mantin, Malaysia — Olutobi Gbenga Ayegbusi
Seri Iskander, Malaysia — Mokhtar Awang

Acknowledgments

The editors would like to thank all the members of the local organizing committee who helped organize the 2nd International Conference on Sustainable Design, Engineering, Management, and Sciences (ICSDEMS 2020), which was conducted through a virtual presentation during December 8 and 9, 2020. We would like to thank the colleagues and staff members at the institutions and organizations that served as partners for the international conference. Their support in organizing a successful conference has helped the editors to gather ideas and papers presented in this book. The editors are grateful to all the speakers who attended the conference and shared from their wealth of experience some exciting findings who have further propelled us to publish this book.

The editors also appreciate various people, including the production team at Springer, who helped and contributed to the creation of this book. We thank all the authors and contributors who presented at the conference and sent us their papers for peer review. The editors would like to thank and appreciate the peer reviewers for their suggestions, comments, efforts, and time spent to go over all the papers.

The editors appreciate the support of the leadership team of their respective institutions for the support, encouragement, and enabling environment created to prepare this book. The conference has inspired and brought the editors together from different disciplines, institutions across different countries, and continents of the world to work on this book. The creation of this book has helped us to become a formidable team. The process has been enjoyable, challenging, inspiring, and more peaceful than we ever thought. We thank you all!

Contents

The Comparison of the Best Practices of the Community-Based Education for Living Heritage Site Conservation

Noor Azramalina Abdul Aziz, Noor Fazamimah Mohd Ariffin, Nor Atiah Ismail, Sumarni Ismail, Anuar Alias, and Nangkula Utaberta

Abstract By a lack of support, appreciation, and understanding, many expressions and manifestations of intangible cultural heritage are under threat, endangered by globalization and cultural homogenization. Intangible cultural heritage has risks becoming lost forever or frozen as a practice belonging to the past if not raised in the community. To strengthen the intangible cultural heritage keeps it alive and continuity, preserving this heritage and passing it on to future generations, and allowing it to change and adapt to any situation. Quality education as a Sustainable Development Goal number 4 (SGD 4) has identified a critical priority in investigating the role of intangible cultural heritage in expanding Education for Sustainable Development. Safeguarding intangible cultural heritage can effectively contribute to sustainable development within each economic, social, and environment while also contributing to peace and security. This paper provided the literature review as an analysis method with a rationalistic approached and discussed the Living Heritage Site Conservation Education for the Community-based in Penang (Malaysia), Singapore, Philippine, and Europe. It is important to note that

N. A. A. Aziz · N. F. M. Ariffin (✉) · N. A. Ismail
Department of Landscape Architecture, Faculty of Design and Architecture, Universiti Putra Malaysia, 43400 UPM Serdang, Selangor Darul Ehsan, Malaysia
e-mail: fazamimah@upm.edu.my

N. A. Ismail
e-mail: natiah@upm.edu.my

S. Ismail
Department of Architecture, Faculty of Design and Architecture, Universiti Putra Malaysia, 43400 UPM Serdang, Selangor Darul Ehsan, Malaysia
e-mail: sumarni@upm.edu.my

A. Alias
Department of Estate Management, Faculty of Built Environment, Universiti Malaya, Wilayah Persekutuan Kuala Lumpur, Jalan Universiti, 50603 Kuala Lumpur, Malaysia
e-mail: anuar_a@um.edu.my

N. Utaberta
Department of Architecture, Faculty of Civil Engineering and Built Environment, Universiti TunHussein Onn Malaysia, Batu Pahat, Johor, Malaysia
e-mail: nangkula@uthm.edu.my

L. Yola et al. (eds.), *Sustainable Architecture and Building Environment*, Lecture Notes in Civil Engineering 161,
https://doi.org/10.1007/978-981-16-2329-5_1

any efforts to conserve the intangible cultural heritage should be aimed not only at development benefits; the most important thing is to comprehend the local people's needs. It is indispensable to promote public awareness of intangible cultural heritage conservation protection toward sustainable development using education programs to achieve cultural heritage conservation efficiency. In conclusion, the key a quality education, which brought in the Community-based Education for Living Heritage Conservation, fulfills people's needs and helps link and empower both local people and the community to conserve the living heritage successfully.

Keywords Intangible cultural heritage · Non-formal education · Public awareness · Quality education · Sustainable Development

1 Introduction

Preservation is defined as an exertion to sustain the heritage elements in both tangible and intangible elements—the tangible cultural heritage such as buildings, landscapes, structures, sites, and communities and, the intangible elements, including oral tradition, music, and cultural activities [1, 2]. In 1960, R. Williams, an influential thinker of the twentieth century, recorded that it could not be compress culture as tangible products because it continuously exists and evolves. He caught the cultural heritage of the essential elements defining human communities' living culture, evolution, and continuing development. Cultural heritage as a culture of a people includes all unnecessary elements considered by a given community as compulsory components of its intrinsic identity and its uniqueness and distinctiveness compared to all other human groups showing the very heart of its distinctive trait.

UNESCO launched the futures of education initiative to reform how knowledge and learning can build humanity's fortunes in a context of rising complexity, unpredictability, and modification in September 2019. UNESCO wants to react to challenges from climate change to growing inequalities, artificial intelligence to learning outcomes, and opportunities that will impact future generations' education and well-being. Now, UNESCO asked people worldwide to contribute their perspective on what the future of knowledge can be. The purpose of the futures education initiative looks to understand how education can build the future of humanity and the planet in 2050 and beyond. The effort is mobilizing a global debate worldwide to involve youth, educators, civil society, governments, businesses, and other stakeholders on how knowledge, teaching, and learning need to reform to address today's and tomorrow's challenges. This initiative uses the concept of futures to identify the rich diversity of ways of knowing and being around the world.

Intangible cultural heritage (ICH) showed the variety of living heritage of humanity and the most influential cultural diversity tool. ICH's primary 'constitutive factors' are determined by ICH's 'self-identification' as an imperative element of its creators and bearers' cultural identity. By its connection with the communities and groups' cultural identity, by its authenticity, and by its indissoluble relationship with human rights, ICH fixed recreation in feedback to the historical and social transformation of the communities and groups concerned [3]. Place identity connected to definition and perception believed by the people regarding their environment, and the loss of identity weakens the depth of meaning, attachment, and diversity of place experience. These evident that new developments within the city center's traditional settings transform build places and place definitions and extensions entrenched in the existing social and cultural context [4].

This paper proposes analyzing the best non-formal education initiative and recognizing the significance of living heritage site conservation for the community toward sustainable development from literature reviews. This paper's objective focused on strengthening, preserving, and passing the intangible cultural heritage that keeps its continuity on to future generations in the community.

2 Methodology

This review article identified the non-formal education initiative for living heritage site conservation completed at Penang (Malaysia), Singapore, Philippine, and Europe. The limitation of this study analyzed the best non-formal education initiative truth activities and events of living heritage. At the same time, it's focused on educating a quality education of the living heritage of the adult and young generations in the community.

3 Findings and Discussions

This study focuses on The Comparison of the Best Practices of the Community-based Education for Living Heritage Site Conservation in Cultural Heritage Education Program in Penang (Malaysia), Singapore, Philippine, and Europe. Table 1 summarizes Cultural Heritage Education Program (CHEP) for the Community-based in Penang (Malaysia), Singapore, Philippine, and Europe and discusses the Establishment, Focus Group, Purpose, and Learning Content of Cultural Heritage Education.

Table 1 showed the CHEP's establishment in Penang, Malaysia, collaborates between Penang government, George Town World Heritage Incorporated, with NGO, Penang Education Association (Arts-ED). This program was launched on 19th November 2016. Penang Education Association (Arts-ED) was the partner of choice for community-based arts and creative education in rural and urban communities [5]. Arts-ED collaborates with local institutions, arts educators, artists, cultural activists, and community associations. Meanwhile, in Singapore, the establishment of the CHEP just in government. The Singapore National Heritage Board (NHB), together with its museums, heritage institutions, and divisions, will implement this program. The National Heritage Board (NHB) was formed on 1st

Table 1 The summary of the cultural heritage education program for the community-based in Penang, Singapore, Philippine, and Europe

	Penang (Malaysia)	Singapore	Philippine	Europe
Establishment	George Town World Heritage Incorporated, in collaboration with non-profit organization Persatuan Pendidikan Pulau Pinang (Arts-ED)	The Singapore National Heritage Board (NHB), together with its museums, heritage institutions and divisions	The National Commission for Culture and the Arts (NCCA), particularly of the Sub commission on Traditional Arts and Cultural Communities (SCCTA)	The Association of Cultural Heritage Education in Finland in partnership with San Millán de la Cogolla Foundation in Spain, the Centre for Urban Development in Serbia, and VITECO E-learning solutions in Italy
Focus Group	Young people aged 10 to 17	- Pre-school and Primary - Secondary and Tertiary - All ages	Transfer of indigenous knowledge and skills to the young people by the cultural masters in the community	10–16-year-old children and young people from Finland, Serbia and Spain
Purpose	To engage young individuals, the local community, educators, and government agencies on the importance of cultural sustainability in the face of twenty first-century challenges	To learn about Singapore's culture and heritage at various stages of the academic journey	To promote the 2003 convention and its implementation through the project proposed by the International Research Centre for Intangible Cultural Heritage in the Asia–Pacific Region (IRCI), i.e., one of the category two centers of UNESCO	To support and make possible transnational learning from and about cultural heritage and allow children and young people to define and voice what they regard as meaningful cultural heritage

(continued)

Table 1 (continued)

	Penang (Malaysia)	Singapore	Philippine	Europe
Learning Content	Learn about tangible and intangible cultural heritage	Learn about tangible and intangible cultural heritage using the life cycle approach to understand the Humanities, Social Studies, Character and Citizenship Education, and twenty first Century Competencies	Learn about five board 'domains' in Intangible Cultural Heritage from the learning Modules for a School of Living Traditions on Buklog Thanksgiving Ritual of the Subanen	Learn about tangible and intangible cultural heritage in museums, heritage centers, and various informal education contexts, as well as on non-formal learning

August 1993 [6]. As the custodian of Singapore's heritage, NHB is responsible for telling the Singapore Story, sharing the Singaporean experience, and imparting Singapore spirit. In Philippine, the CHEP or called School Living Traditional (SLT) is a flagship program of the National Commission for Culture and the Arts (NCCA), particularly of the Subcommission on Traditional Arts and Cultural Communities (SCCTA), conceived to safeguard traditional cultural practices of indigenous peoples [7]. Several community-managed and NCCA-supported SLTS have been established in several parts of the Subanen traditional home range since 2012. In Europe, the CHEP or called Heritage Hubs was a two-year project related to the European Year of Cultural Heritage 2018, co-funded by the Creative Europe program of the European Union. This project to enhance the role of heritage culture within young people ordinary life thought of as a vehicle to drive a sense of belonging to the common land such as the European Union. The project is coordinated by the Association of Cultural Heritage Education in Finland in partnership with the San Millán de la Cogolla Foundation in Spain, the Centre for Urban Development in Serbia, and VITECO E-learning solutions in Italy [8]. The organization's CHEP establishment as a trace and base of conservation in tangible and intangible cultural heritage in Table 1. People of expertise from different fields of cultural heritage join together to provide practical education in this program. From this establishment, the mission, vision, and purpose of CHEP will be created as a guide to achieve successful education.

The CHEP focus group in Penang, where young people aged 10 to 17 are primary and secondary school students [5]. It's the same with the CHEPs focus group in Europe aged 10 to 16-year-old children and young people from Finland, Serbia, and Spain [8]. Meanwhile, in Singapore, the CHEPs focus group has three-level that (1) Pre-school and Primary, (2) Secondary and Tertiary, and (3) All ages. As part of the cultural heritage education using the life cycle approach, these programs will accompany every individual right from their pre-school days to tertiary education level to learn about Singapore's culture and heritage at various stages of the academic journey [6]. Different in the Philippine, the CHEPs focus group on the transfer of indigenous knowledge and skills to the young people by the cultural masters in the community [7]. The Subanen indigenous community in the Philippine use this opportunity to pass down their knowledge and skills to the young generation to implement and practice in their daily life. In these four case studies, all the CHEPs focus group focused on transmission knowledge and skills to the young generation for cultural heritage continuity represented in Table 1. But it lacks in adult cultural heritage education. It's necessary to remind adults about their cultural heritage because not all individuals get cultural heritage education from a young age. If this intangible cultural heritage is not maintained, it caused losses to the community representing an image, identity, sense of belonging, and community pride.

According to George Town World Heritage Incorporated and Penang Education Association (2018), the CHEP's purpose in Penang to engage young individuals, the local community, educators, and government agencies on the importance of cultural sustainability in the face of twenty first-century challenges. Engaging and well-curated, the activities enable the younger generation to appreciate, understand, and interpret the George Town UNESCO World Heritage Site through different participation levels. In Singapore, the CHEP aims to learn about Singapore's culture and heritage at various academic journeys [6]. The key competencies necessary to thrive in the twenty-first century of the CHEP in Singapore, the education partners, and schools were to ensure that content and programming remain at the cutting edge of heritage education while complementing the Ministry of Education's efforts to provide a holistic education. Meanwhile, the purpose of the CHEP in the Philippine to promote the 2003 convention and its implementation through the project proposed by the International Research Centre for Intangible Cultural Heritage in the Asia–Pacific Region (IRCI), i.e., one of the categories two centres of UNESCO [7]. And in Europe, the CHEP's purpose is to support and make possible transnational learning from and about cultural heritage and allow children and young people to define and voice what they regard as meaningful cultural heritage [8]. In Table 1, the CHEP's purpose in these four case studies showed that the cultural heritage was essential to the community because it contributes to the continuity between the past, present, and future. The knowledge of history is vital to understand the current societies, based in turn on future construction.

In Penang, the learning content of the CHEP about tangible and intangible cultural heritage. According to George Town World Heritage Incorporated and Penang Education Association (2018), from the activity of Traditional Games (Wa Wa Warisan), Thematic Heritage Exploration Trails, Traditional Craft (School of Craft), Traditional Food and Local Ingredients (You Think You Can Masak), and Creative Arts Workshop based on a Cultural Heritage Site (Youth Arts Camp). The young generation can learn about cultural heritage differences in multiculturalism at Georgetown, Penang. The CHEP's Learning content in Singapore is also about tangible and intangible cultural heritage using a life cycle approach to understanding the Humanities, Social Studies, Character and Citizenship Education, and twenty first Century Competencies [6]. Same with the CHEP in Europe, the learning content about tangible and intangible cultural heritage. The museums, heritage centers, and various informal education contexts and non-formal learning, such as social encounters and intergenerational learning in families, combine virtual and real-life experiences and heritage interpretation. Various digital, classroom and outdoor learning options are used to learn about tangible and intangible cultural heritage in the Cultural Heritage Education Program in Europe [8]. Meanwhile, in the Philippine, the CHEPs learning content focused on intangible cultural heritage. The Learning Modules for a School of Living Traditions on Buklog Thanksgiving Ritual of the Subanen was provided based on five board 'domains' in Intangible Cultural Heritage [7]. In these four case studies, three case studies in Penang, Singapore, and Europe, the CHEPs learning content about tangible and intangible cultural heritage just one case study in Philippine the CHEPs learning content about intangible cultural heritage. In Table 1 showed that the conservation of tangible and intangible cultural heritage was important in living heritage sites to maintain cultural heritage continuity. The heritage places continue to be used for the purpose that has initially been built or has acquired new functions or use. New functions may be touristic, economic, or social such as converting buildings to museums [9]. When we were preserving the heritage building or monument, we also need to safeguard our community's traditional cultural practices as a living heritage site. Traditional cultural practices in the community are essential in our daily life to continuity our image, identity, sense of belonging, and community pride.

4 Conclusion

In conclusion, the key a quality education of living heritage education conservation for the community not only fulfills people's needs but also helps link and empower both local people and the community to conserve the living heritage effectively and at the same time increase public awareness and participation. This study provides the Community-Based Education (CBE) Framework for Living Heritage Site Conservation and the Community Participation Level toward Sustainable Development at the World Heritage Site in Malaysia as the new direction for further research. The adult cultural heritage education program in community-based also needs further investigation.

Acknowledgements Fundamental Research Grant Scheme supported this study under the Ministry of Higher Education Malaysia (FRGS/1/2019/WAB04/UPM/02/2). The authors would like to thank University Putra Malaysia for their support.

References

1. Hani U et al (2012) Preserving cultural heritage through creative industry: a lesson from SaungAngklungUdjo. Proc Econ Financ 4:193–200
2. Mohd Baroldin N, Mohd Din SA (2018) Conservation planning guidelines and design of Melaka heritage shophouses. Asian J Environ Behav Stud (ajE–Bs), 3(8):61–70 (May/Jun 2018)
3. Lenzerini F (2011) Intangible cultural heritage: the living culture of peoples. Eur J Int Law 22(1):101–120. https://doi.org/10.1093/ejil/chr006
4. Ujang N (2017) Place attachment and continuity of urban place identity. Asian J Environ Behav Stud ajE–Bs 2(2):117–132
5. George Town World Heritage Incorporated and Penang Education Association (2018) Cultural heritage education program (CHEP) Booklet 2018. https://issuu.com/arts-ed.penang/docs/chep_booklet_april_2_2018
6. Singapore National Heritage Board (2020) Heritage education highlights 2020. https://www.nhb.gov.sg/-/media/nhb/images/education/resources-for-school/heritage_education_highlights_2020.pdf?la=en
7. The International Research Centre for ICH in Asia-Pacific Region, under the auspices of UNESCO, and the National Commission for Culture and the Arts (2019). Multi-disciplinary study on intangible cultural heritage's contribution to sustainable development, focusing on education: a guide for facilitators and local coordinators for a school a living traditions on Buklog of thanksgiving ritual of the Subanen, Philippines.
8. The Creative Europe Program of the European Union (2020) Heritage hubs: manual for cultural heritage education. https://heritagehubs.eu/wp-content/uploads/2020/03/HH-Manual-Inglese-Web-rev2.pdf
9. ICCROM (2013) Living heritage: a summary. https://www.iccrom.org/wp-content/uploads/PCA_Annexe-1.pdf

Consequences of Under-Reporting of Occupational Accidents in the Nigerian Construction Industry

Kamoli Adetunji, Adul Hamid Razali, and Syamsul Hendra Bin Mahmud

Abstract An essential and major part of each prevention policy is to have explicit information on the figure of diseases and accidents, their seriousness, the origins, the work environment, and sectors where they happen. This means that every organization and nation required a well-active and efficient system for recording and reporting diseases and accidents. This will serve as a source of information for the investigation, policy formulation, and prevention of incidents in the workplace. Inadequate resources and bad reporting mechanisms could support the under-reporting of occupational accidents. The study's aim is therefore to examine the consequences of under-reporting of occupational accidents in the Nigerian construction industry. A review of the literature was conducted on under-reporting of occupation accidents followed by a field survey with 300 structured questionnaires, administered to solicit information from construction professionals in the Federal Capital Territory, Abuja, Nigeria. The SPSS software and Microsoft excel were used to analyse the data. The findings of the study revealed the consequences of under-reporting among which are affecting the implementation policy on occupational health and safety, and equally affecting the planning and prevention of occupational illness and accidents. The study, therefore, suggests an improvement in any existing reporting structure and also establishes an unsophisticated and robust system of reporting to motivate construction organizations and employees to account for the full detail of all incidents that occur in the working environments.

Keywords Under-reporting · Occupational accident · Health · Safety · Construction industry

K. Adetunji (✉) · A. H. Razali · S. H. Bin Mahmud
Department of Quantity Surveying, Universiti Teknologi Malaysia, Johor Bahru, Malaysia
e-mail: kamoli1972@graduate.utm.my

L. Yola et al. (eds.), *Sustainable Architecture and Building Environment*, Lecture Notes in Civil Engineering 161,
https://doi.org/10.1007/978-981-16-2329-5_2

1 Introduction

The under-reporting of incidents in the organizations could be supported by the demonstration of a poor health and safety culture and insufficient reporting mechanisms put in place for recording and reporting hazardous events and accidents [1]. One serious feature of the ineffective application of the reporting scheme is the lack of management commitment. Management commitment produces a higher degree of motivation for the workers and consideration for the management of health and safety in the organization [2]. Therefore, the active participation of top management in health and safety in the workplace is indispensable. The negative consequences of lack of commitment to health and safety by the management in the organization can have adverse effects on both the organization and employees such as increase absenteeism of employees, loss of production, equipment damages, and a bad image of the organization as a result of higher numbers of accidents occurred in the organization [3].

Accurate reporting of occupational illnesses and injuries is essential to observe workplace safety and health and to recognize the most desirable interventions for the prevention of occupational accidents [4, 5]. It was reported that there was a discrepancy in the figure of accidents reported to the occupational health and safety department of the federal ministry of labour and employment and that of the Nigerian social insurance trust fund [6]. This has a lot of negative implications on health and safety management in Nigeria. Therefore, the study was conducted to identify the consequences of under-reporting of occupational accidents in the Nigerian construction industry.

2 Literature Review

The majority of the accidents experienced in the workplace go unrecorded with the figure of individual under-reporting varying from 71 to 80% [7, 8]. It was also revealed that the underestimate rate of work-associated illness, injuries, and the level of fatalities could be as far as a hundred percent [4]. Nayanthara and Uthpala [9] revealed that 80% of construction incidents are under-reported. Some studies had associated accident under-reporting at the employees level due to age [10], fear of deprivation of benefits [11], and the perception that injuries are part of work and nothing can be done to stop it [4]. Employees are less likely to report accidents in the workplace when unemployment is high [12]. Similarly, establishment under-reporting has been associated with issues like size [13] sector [14] absence of management understanding of the importance of accidents information [15], and establishments' health and safety climate [16]. The notions that organizations could be subjected to sanction for high records of occupational accidents could be among the reasons for under-reporting accidents in Nigeria [6]. Also, high accidents records give a bad image to the organization, therefore, contractors records of accidents forward to the appropriate authorities are likely to be imprecise [17].

Furthermore, the quantity and quality of health and safety information in and among teamwork and management might also account for under-reporting. A climate that disheartens reporting health and safety matters with supervisors, co-employees, and management would probably nurture a climate that is encouraging under-reporting workplace accidents at both organizational and individual levels. An organization with a poor health and safety climate is less likely to firmly enforce health and safety policies despite if they are legally stated [16]. Though it is a legal obligation for the organizations to keep precise logs, one can forecast that the implementation of such a policy would be more neglectful in an establishment with a bad health and safety climate. Mandatory reporting systems have been strengthened by automatic logging systems in the railway sector [15]. But, the recording of various cases still relies on individual employees and organizations to complete the appropriate paperwork.

The previous research outlines several issues that impact incident reporting. Powell et al. [18] established that cases were not reported if they were observed as part of the job. Equally, Beale et al. [19] highlighted that the reporting of ferocity occurrences by publicans and established that some level of severity was acknowledged as the norm. Suppression and under-reporting of information have been associated with the presence of a blame philosophy where the drive of assembling incident and accident information is to allot blame rather than take correcting measures [20, 21].

A level of reliance is needed for employees to report cases. Some establishments have applied confidential reporting systems to this end [22]. Whereas others have attempted to design a no blame philosophy that boosts the reporting of cases as a method of improving safety and health rather than apportions blame [23]. Control of accidents is important in the construction industry, organizations are required to appraise the hazards and take appropriate measures to safeguard, ensure the health and safety of their employees, minimizing risk through continuous surveillance and observing where incidents are likely to happen. And more importantly, the organization must keep accurate records of all incidents that occurred in the workplace.

3 Material and Methods

The study aim is to examine the consequences of under-reporting of occupational accidents in the Nigerian construction industry. The literature review of under-reporting of occupational accidents was conducted, then followed by a field survey. 300 questionnaires structured in a 5-point linker scale were administered to the construction professionals in construction and consulting organizations and 235 were returned which represent 78.33%, therefore considered suitable for the analysis of the research. The set of questionnaires consist section that comprises the variables on the consequences of under-reporting of occupational accidents in the

Nigerian construction industry. The data collected were analysed with SPSS software and Microsoft excel. This is to find out the importance of factors that represented the consequences of under-reporting of occupational accidents in the Nigerian construction industry. The factors put in rank, the top-ranked is more significant than the next.

4 Results and Discussion

Table 1 shows the result of the respondents' view on the consequences of the under-reporting of occupational accidents in the Nigerian construction industry. Affecting implementation policy on occupational health and safety and affecting planning and prevention of occupational illness and accidents both have the highest relative importance index of 0.804. Suppressing accurate reporting can result in missed opportunities to identify cases at an early stage has a relative importance index of 0.800 and the result on inadequate budgetary provision for occupational health and safety has a relative importance index. 0.762. The result of the study agrees with the position of [4] which highlighted that management failure to record the injury occurrence statistics could hinder decision making concerning the setting of prevention priorities and for allocating resources for health and safety matters. The authors further emphasized that plant managers must recognize that suppressing accurate reporting could result in missed opportunities to identify conditions at an early phase. Similarly, [24] emphasize that when an organization under-reports occupational injury, its employees' compensation affair will not reveal its real injury experience, in such a situation, the erroneous information could be used for the management of health and safety in the organization. Likewise, if organizations do not recognize entire occupational incidents, they will not properly evaluate the extent and cost of occupational injury and might not target internal health and safety resources adequately. Please read the Instructions to

Table 1 Consequences of under-reporting of occupational accidents

The consequences of under-reporting of occupational illnesses and accidents	Sum	Mean	Std. Dvt	Relative importance index	Ranking
Affecting planning and prevention of occupational illness and accidents	945	4.02	1.084	0.804	1
Affecting implementation policy on occupational health and safety	945	4.02	1.084	0.804	1
Suppressing accurate reporting can result in missed opportunities to identify cases at an early stage	940	4.00	1.169	0.800	3
The result of inadequate budgetary provision for occupational health and safety	895	3.81	1.144	0.762	4

Authors documents carefully. Manuscripts that do not comply will be returned for correction. The following should be noted in particular.

5 Conclusions

The consequences of under-reporting of occupational accidents in the Nigerian construction industry are enormous namely affect the adequate planning and prevention of accidents and illness, affecting the planning and implementation of occupational health and safety policy, result in inadequate budgetary provisions for health and safety and suppressing accurate reporting can result in missed opportunities to identify cases at an early stage. Monitoring and assessment are vital for efficient prevention of under-reporting of occupational accidents, and review of the management system in agreement with the guidelines is equally essential. Effective mechanisms must be put in place for the record-keeping and reporting procedures. This forms the central point of the program. An active monitoring system can be put in place for the evaluation of accidents and incidents in the Nigerian construction industry. Reporting of accidents is important as it is providing appropriate interventions for accidents and illness prevention. The establishment of an uncomplicated and robust reporting system to encourage construction organizations and employees to account for the detail of all accidents is essential. The system should be simple to understand, and less time consuming to use. The application of less formal methods to account for less severe accidents might help to control the difficulty of under-reporting. This can be accompanied by the support of the organizations to report all accidents. It is worthy to note that accurate reporting of occupational accidents can help to recognize unsafe work practices and areas, measure the influence of occupational problems on productivity, safety, health, and costs and appraise the impact of health and safety improvements.

References

1. Rosli, Andy (2011) Accident under reporting in the Malaysian construction. In: EACEF-international conference of civil, pp 216–222
2. HSE (1993) Safety culture. Hse [Internet], pp 1–4. Available from https://www.hse.gov.uk/humanfactors/topics/common4.pdf
3. Randall DM (1987) Commitment and the organization: the organization man revisited. Acad Manag Rev 13(3)
4. Glenn et al (1999) Under-reporting of work-related disorders in the workplace: a case study and review of the literature. Ergonomics 42(1):171–182
5. Boden LI, Ozonoff A (2008) Capture-recapture estimates of nonfatal workplace injuries and illnesses. Ann Epidemiol 18(6):500–506
6. Federal ministry of labour and employment (2016) Nigeria country profile on occupational safety and health. Federal ministry of labour and employment

7. Tahira and Armando (2010) Accident under-reporting among employees: testing the moderating influence of psychological safety climate and supervisor enforcement of safety practices. Accid Anal Prev [Internet] 42(5):1438–1444. Available from https://doi.org/10.1016/j.aap.2009.06.027
8. Tahira and Maja (2013) Pressure to produce = pressure to reduce accident reporting? Accid Anal Prev [Internet] 59:580–587. Available from https://doi.org/10.1016/j.aap.2013.07.020
9. Nayanthara, Uthpala (2018) Under-reporting of construction accidents in Sri Lanka. J Eng Des Technol 16(6):850–868
10. Hugh, Jens (1998) Occupational injury and illness rates, 1992–96: why they fell. US Dep Labor [Internet] 121(September):1–550. Available from https://www.bls.gov/iif/oiics_manual_2010.pdf
11. Robert, Lois (2003) Pay and benefits: the role of compensation systems in workplace safety. Psychol Work Safety
12. Palali A, van Ours JC (2017) Workplace accidents and workplace safety: on under-reporting and temporary jobs. Fond Giacomo Brodolini John Wiley Sons Ltd, vol 31, issue 1, pp 1–14
13. Leigh et al (2004) An estimate of the U.S. Government's undercount of nonfatal occupational injuries. J Occup Environ Med 46(1):10–8
14. Daniels C, Marlow P (2005) Literature review on the reporting of workplace injury trends HSL/2005/36. Health and safety laboratory Harpur Hill, UK
15. Clarke S (1998) Organizational factors affecting the incident reporting of train drivers. Work Stress 12(1):6–16
16. Probst TM, Brubaker TL, Barsotti A (2008) Organizational injury rate underreporting: the moderating effect of organizational safety climate. J Appl Psychol 93(5):1147–1154
17. Idoro GI (2010) Management health and safety management efforts as correlates of performance in the Nigerian construction industry. J Civ Eng Manag 14(4):277–285
18. Powell et al (1971) 2000 accidents: a shop floor study of their causes
19. Beale et al (1994) The role of the reporting of violent incidents in tackling workplace violence. In: Proceedings of the fourth conference on safety and well-being, lough-borough, November (Loughborough), pp 138–151
20. Adams, Hartwell (1977) Accident-reporting systems: a basic problem area in industrial society. J Occup Psychol 50(4):285–298
21. Webb et al (1989) Filtering effects in reporting work injuries. Accid Anal Prev 21(2):115–123
22. Reynard WD (1986) The development of the NASA aviation safety reporting system: volume 1114 of NASA reference publication. National Aeronautics and Space Administration, pp 1–204
23. Turner BA (1991) The development of a safety culture. Chem Ind 241–243
24. Glazner et al (1998) Construction injury rates may exceed national estimates: evidence from the construction of Denver International Airport. Am J Ind Med 34(2):105–112

Having More Access Leads to Better Welfare: Study of Access for People with Disabilities (PwD) and Its Spatial Impact

Yusuf Faisal Martak and Renny Nurhasana

Abstract Disability groups are included in vulnerable groups who need special assistance to improve their quality of life. One of the most important things needed for persons with disabilities is the existence and ease of access to various sectors such as health and education. This study aims to see how the influence of the existence of disabilities on welfare and how the existence of access can improve the welfare of persons with disabilities in the region and in the surrounding area. Using the OLS regression model at the household and district/city levels, the result shows that to date people with disabilities are still a negative factor for welfare and the existence of access to health insurance, education and health facilities for disabilities can improve the welfare of their group. Meanwhile, the Spatial Autoregressive Model shows the existence of access to good health and education facilities in an area can improve the welfare of persons with disabilities in the surrounding area that is close to their home area. Whereas, the impact for other areas as a whole only applies to health facilities and access *to health insurance.*

Keywords Disability · Disability Access · Spillover · Spatial · Impact

1 Introduction

According to Indonesia Law No. 8/2016, persons with disabilities are people who have various types of limitations such as physical and non-physical limitations. Non-physical limitations are divided into several types, namely mental, sensory, and intellectual. Meanwhile, according to Indonesia Law No. 39/1999, groups of people with disabilities are one of the groups entitled to receive more treatment and protection than non-disabled groups, especially in the education, health, and employment sectors.

Y. F. Martak (✉) · R. Nurhasana
Urban Studies Program, School of Strategic and Global Studies, Universitas Indonesia, Central Jakarta 10430, Indonesia

L. Yola et al. (eds.), *Sustainable Architecture and Building Environment*, Lecture Notes in Civil Engineering 161,
https://doi.org/10.1007/978-981-16-2329-5_2

Based on data from the National Socio-Economic Survey (SUSENAS) 2019, the percentage of persons with disabilities in Indonesia is in the range of 9–10% of the total population. The large percentage indicates that persons with disabilities have a significant role in the economy and have the opportunity to help improve national welfare. However, limitations for persons with disabilities to carry out economic activities even non-economy make the contribution of this group not optimal [10]. In addition, these limitations also bring disabilities and families who have members with disabilities into lower economic status compared to non-disabled people. According to research conducted by Bella and Dartanto [1] in Indonesia, a family whose head of household is a person with a disability has the opportunity to become less prosperous. Moreover, not only the head of the household, everyone who falls into the group with disabilities has a lower life indicator than non-disabled people [7]. This explains that families with household members with disabilities regardless of their role, are still unable to independently promote the welfare of their families which results in inequality in welfare between groups with disabilities and non-disabled people. One of the reasons for the emergence of this incident is the limitation of movement or space for persons with disabilities [7]. To increase the contribution of groups with disabilities and reduce existing inequalities, one thing that can be done is to provide access for groups with disabilities by the government to various sectors that have been previously mentioned [10].

Access for persons with disabilities in Indonesia for each region is still very diverse, this is due to differences in regional focuses and priorities in directing policies towards increasing access for persons with disabilities. Table 1 shows a picture of the unequal availability of access for persons with disabilities in Indonesia. Focusing on access to the education sector, based on 2019 SUSENAS data, the difference between the highest and lowest disability participation rates at the district/city level is 87%. Regarding access to health, the difference in the ratio of disabilities with the highest and lowest health insurance coverage is 76.28%, while in the calculation of the number of disabilities per one hospital, the highest area is 65,425 disabilities per hospital and the lowest is 660. Still, in the same source, it can be shown that the difference in regions based on the ratio of disabilities who have jobs is 74.63%. Of these four indicators, regional disparities in managing access for disabilities are still very diverse, and this provides different benefits according to the availability of access they have.

Table 1. Description of access inequality for disabilities in the regions based on education, health, and employment indicators

	School participation rate (%)	Health insurance ownership ratio (%)	Job ownership ratio (%)	Disabilities per one hospital
Lowest	13	23.72	25.37	660
The highest	100	100	100	65,425
Difference	87	76.28	74.63	64,765

Source SUSENAS 2019, compiled

The purpose of this study is to measure how the current contribution of disabilities to the national economy and what factors can improve the welfare of persons with disabilities at the household, regional, and other regional levels. Based on these objectives, the following analysis will be carried out: (1) the impact of the existence of disabilities on household and regional welfare; (2) the impact of having access to education, health, and employment for disabilities on household and regional welfare; (3) spillover effect of the impact of the presence of disabilities in one area on other areas.

2 Data and Methodology

2.1 Data Sources & Method of Analysis

The scope of observation in this study is Indonesian country. The data used in this study are individual and village data from the National Socio-Economic Survey (SUSENAS) KOR March 2019 and Village Potential (PODES) 2018. SUSENAS and Village Potential are data sourced from the Central Statistics Agency (BPS), where the survey was conducted twice a year for SUSENAS and once every 4 years for Village Potential. In general, SUSENAS data is the main data source used in this study. Several variables are used as proxies for socio-economic, education, health, and several other indicators. This study uses the Village Potential data variable in the number of hospitals in the area. Referring to these data, the method used in this study is an empirical quantitative model divided into household and districts level explained in each section. Moreover, this study also tries to describe how the social impact of having access to disabilities in one area on the welfare of other areas.

2.1.1 Household Level

An empirical model at the household level is carried out to see how disability contributes to household welfare, as well as what factors can affect the welfare of households with disabilities themselves. The household-level empirical model uses an econometric approach, namely Ordinary Least Square (OLS) regression. OLS is a simple regression model that is generally used to measure how one variable changes to another [12]. The OLS equation used in the study is as follows:

$$\ln \text{percapita expenditure}_i = \beta_0 + \beta_0 + \beta_1 D.disability_i + \beta_n Control\_variable_i + \varepsilon_i \tag{1}$$

The dependent variable of the OLS equation is the natural logarithm of per capita household expenditure, the use of natural logarithms is intended to reduce the variance of per capita household expenditure. Meanwhile, the main independent variable used was the presence of disabilities in the family, followed by several other control variables.

After seeing how the impact of the existence of disabilities on household welfare, equations are then formed to see what factors can improve the welfare of families with disabilities. This is intended to analyze the main factors that can improve the welfare of families with disabilities. Still in the OLS and logistic regression methods, the equation used is as follows:

$$\begin{aligned} &\text{In}\, percapita\ expenditure\ (for\ HH\ with\ Disabilities)_i \\ &= \beta_0 + \beta_n Disability\, access + \beta_n Control\_variable_i + \varepsilon_i \end{aligned} \quad (2)$$

2.1.2 District Level

Apart from conducting analysis at the household level, this study also attempts to conduct analysis at the district/city level. The purpose of this analysis is to see how the role of disabilities in the overall welfare of the region and what factors can improve the welfare of disabilities from a regional perspective, in this case, is the district/city level. This was done using SUSENAS 2019 data juxtaposed with Village Potential 2018.

There are 4 indicators used to measure access for persons with disabilities in the regions: (1) working disability ratio; (2) high availability of health insurance for persons with disabilities; (3) the existence of health facilities compared to the high number of persons with disabilities; (4) high school participation rates for persons with disabilities. Some of these indicators have represented access to disabilities from various sectors, such as the health, education, and employment sectors.

2.1.3 Spillover Effect

In addition to regression analysis at the household and regional level, this study also seeks to see how the effect of having access to persons with disabilities affects the welfare of disabilities in other areas. In conducting this analysis, the most appropriate model to use is the Spatial Autoregressive Model [4]. The purpose of the SAR model in general is to see the spatial impact of exogenous/independent variables on a dependent variable in other areas [8]. In the SAR model, the equation used is as follows:

$$Outcome_i = \beta_0 + \beta_1 X_i + \gamma 1 \sum_{j=1}^{N} Wij * X + \varepsilon_i \quad (3)$$

There is a big difference between the linear regression equation and SAR, namely in the existence of the spatial matrix itself. The notion of a spatial matrix or what is generally called a weighting matrix is the separation of the places of each individual who are in different places. This is done so that each individual can see how much influence s/he has on other individuals [9]. The following is an illustration of the matrix that will be used.

Table 2 shows an illustration of the spatial matrix in 4 regions, where the sign "*" is the magnitude of the influence of one area on another. In this case, area B is indicated to affect region A, even for C and D. In the spatial analysis, there is no difference with the Eq. 3 analysis (Table 3).

Table 2 Spatial Matrix llustration

	Area A	Area B	Area C	Area D
Area A	0	*	0	0
Area B	*	0	0	0
Area C	0	0	0	*
Area D	0	0	*	0

Source Author

Table 3 Variables and models used in the analysis

Variable	1st equation: disability to HH welfare	2nd equation: disability access to HH disabilities welfare	3rd equation: disability to district welfare	4th equation: disability access to disability district welfare	5th & 6th equation: local and global spatial analysis
Variables represent disability					
Household with disabilities person	Yes				
Ratio of disability/ population			Yes		
Variables represent the access of disability					
Disability work (1 = yes)		Yes			
Disability attending school (1 = yes)		Yes			
Disability healthcare (1 = yes)		Yes			

(continued)

Table 3 (continued)

Variable	1st equation: disability to HH welfare	2nd equation: disability access to HH disabilities welfare	3rd equation: disability to district welfare	4th equation: disability access to disability district welfare	5th & 6th equation: local and global spatial analysis
Ratio of working disability				Yes	
High ratio of school disabilities participation (1 = yes)				Yes	Yes
High ratio of having healthcare disability (1 = yes)				Yes	Yes
High number of disabilities on 1 health facilities (1 = yes)				Yes	Yes
Control variables (in ratio to population if district level)					
Average year schooling	Yes	Yes	Yes	Yes	Yes
Working	Yes		Yes		
Electrification	Yes	Yes	Yes	Yes	Yes
Healthcare	Yes		Yes		
Health facilities			Yes		
School participation disability			Yes		
Urban	Yes	Yes			
Household number	Yes	Yes			

Source Author

3 Result and Discussion

The results section of the analysis and discussion will be divided into 3 parts, namely the impact of disability on households and regions, the existence of disability access to the welfare of the group, and the impact of the existence of disability access for other areas.

3.1 *The Existence of Disabilities on Household and Regional Welfare*

In models 1 and 3 with the aim of seeing the effect of disability on total welfare, the results show that in 2019, the existence of disabilities in Indonesia still has a negative impact on total welfare. In Researchers of [1] also stated that the existence of disability in Indonesian household reduces the level of family welfare, it can also increase the chances of the family falling into the poor category. Referring to the estimation results for 2012 and 2019 which have not changed, this confirms that the negative impact of the existence of disabilities has yet to be resolved or reduced by the policies that have been established. In the case of Indonesia, one of the reason the implementation of disabilities regulation not implemented optimally because the lack of information about what disabilities need most [3]. Some of other reasons why the changes did not occur are that even though the policy was established, the policy did not have a large focus on tackling the groups with disabilities problem [2]. In addition, the existence of access that does not support mental and physical disabilities is another cause of one of the things that makes the lives of persons with disabilities lower, in education and employment as well [10, 11] (Table. 4).

3.2 *Availability of Access for Persons with Disabilities to the Welfare of Their Group*

In models 2 and 4, an analysis is carried out to see how the impact of the availability of access on the welfare of groups with disabilities. The results of the analysis show that at the household level, disabilities who have a job can significantly increase the average welfare of their household. Apart from working, disabilities who have health insurance and attend school also have an equally positive and significant contribution to the welfare of households with disabilities. Not different from household results, results at the regional level also show the same thing. The average expenditure per capita of disabilities in a district/city can be increased significantly if the region experiences: (1) an increase in the ratio of persons with disabilities to work; (2) has a disability health insurance ownership ratio that is above the average; (3) has above average school participation rates for

Table 4 Regression results for the 6 models used

Variable	1st equation: HH level	2nd equation: HH level	3rd equation: District level	4th equation: District level	5th equation: Local spillover	6th equation: Global spillover
Variables represent disability						
Household with disabilities person	−0.034*** (0.002)					
Ratio of disability/ population			−1.708***			
			(0.310)			
Variables represent the access of disability						
Disability work (1 = yes)		0.030***				
		(0.002)				
Disability attending school (1 = yes)		0.050***				
		(0.008)				
Disability healthcare (1 = yes)		0.051***				
		(0.040)				
Ratio of working disability				0.210*		
				(0.131)		
High ratio of school disabilities participation (1 = yes)				0.040** (0.020)	0.102*** (0.037)	0.0001 (0.000)
High ratio of having healthcare disability (1 = yes)				0.067*** (0.023)	0.031 (0.032)	0.0002* (0.000)

(continued)

Table 4 (continued)

Variable	1st equation: HH level	2nd equation: HH level	3rd equation: District level	4th equation: District level	5th equation: Local spillover	6th equation: Global spillover
High number of disability on 1 health facilities (1 = yes)				−0.087*** (0.024)	−0.082** (0.037)	−0.0003** (0.000)
Control variables (in ratio to population if district level)						
Average year schooling	0.071***	0.072***	0.140***	0.121***	0.014	0.0003**
	(0.000)	(0.000)	(0.012)	(0.015)	(0.019)	(0.000)
Working	0.043***		0.820***			
	(0.001)		(0.1715)			
Electrification	0.119***	0.147***	0.186	0.020	−0.124	−0.0002
	(0.005)	(0.010)	(0.127)	(0.230)	(0.152)	(0.000)
Healthcare			0.154***			
			(0.051)			
Health facilities			−0.000***			
			(0.000)			
School participation disability			0.042			
			(0.058)			
Urban	0.181***	0.150***				
	(0.002)	(0.003)				
Household number	−0.139***	−0.119***				
	(0.000)	(0.000)				
Constant	13,560***	13,444***	12,284***	12,559***		
	(0.006)	(0.004)	(0.139)	(0.259)		
R-square	0.33	0.29	0.55	0.37	0.39	0.37
Number of observations	315,672	87,543	496	514	492	492

Robust standard errors in parentheses *** $p < 0.01$, ** $p < 0.05$, * $p < 0.1$

persons with disabilities; (4) has a low ratio of disability rates in one hospital. Through ownership of access proxies for persons with disabilities, based on the above estimates, each region will experience an increase in the welfare of the disabled group.

The existence of access for persons with disabilities in Indonesia is indeed the best solution to improve the welfare of these vulnerable groups. According to Researchers of [6], the absence of access for persons with disabilities can strengthen the vulnerability of these groups. Moreover, the establishment of easy access for persons with disabilities to the employment, education, and infrastructure sectors has a good impact on welfare. In addition, improvisation in access to health can also help reduce the vulnerability of groups with disabilities [10].

3.3 Availability of Access for Persons with Disabilities to the Welfare of Their Groups in Other Areas

Models 5 and 6 show that accessibility for persons with disabilities also has a good impact on the welfare of their groups in the surrounding area. Before the spatial analysis is carried out, there are stages to test whether the variables used have an influence on other areas with the Moran test [5]. Using the Moran test, it is found that there is a strong influence of the area of origin variables on the surrounding area. In the analysis of areas that fall into the category of neighbors of origin, the existence of qualified health facilities and access to education for persons with disabilities can improve the welfare of persons with disabilities in other areas, but not for having high health insurance. Whereas in the analysis of other regions as a whole, only access to health insurance and health facilities can improve the welfare of disabilities in other regions even though the magnitude of the effect is very small. Based on these results, it can be concluded that the existence of access for persons with disabilities has a spillover effect that can improve the average welfare of persons with disabilities in other areas. According to Researchers of [10], there is indeed a relationship between groups with disabilities and the level of welfare in other areas close to the area of origin. In addition, the existence of access, especially for persons with disabilities in an area that is better described through infrastructure, the presence of health workers, and the status of the capital city, in its findings, can improve the welfare of the disabled group as well as reduce the poverty level gap between groups with disabilities and non-disabilities in other areas.

4 Conclusion

Thus far, the disability group is a vulnerable group that still has several problems, such as the small contribution of disability and even tends to be negative for the welfare of society as a whole or for the group. In 2012, disability has a negative impact on household welfare, besides that this group can also increase the chances of a family entering poverty [1]. In the analysis in 2019, it was found that groups with disabilities still have a negative impact on the welfare of the whole and the group. This shows that there has been no change in the role of the disabled in the economy (Models 1 and 3).

One of the reasons for not changing conditions with disabilities is the absence of adequate access for groups with disabilities in every area where they live. Access for disabilities is reflected in their access to health facilities, educational participation, and ownership of health insurance. Although there are some areas that have this access, the inequality of access for persons with disabilities has not been able to help reduce the negative impact of the existence of groups with disabilities nationally.

In the analysis that has been conducted, it can be proven that the existence of access for persons with disabilities is indeed very important because in the case of Indonesia, access from the health sector and education can improve the welfare of families with disabilities at the regional level. In addition, the existence of access that is above the average in other regions can also improve the welfare of groups with disabilities in neighboring areas and in other areas. To improve the welfare of groups with disabilities so as to reduce the overall negative impact, it is necessary to accelerate policies by the government aimed at providing access for persons with disabilities related to health facilities, health insurance, and educational participation. In detail, several things that the government needs to provide are: (1) improving disability-friendly health facilities especially in areas with lower health facilities and provide health workers with special skills for people with disabilities; (2) acceleration in providing health insurance for all of the people with disabilities; (3) enhancing special education facilities for people with disabilities, make additional quota for disabilities in public school, and improve the quality of teacher for Special Education subjects; (4) in the presence of spillover effect, the government may focus provides access for disabilities in the area with high rates of spillover effect such as metropolitan area, even though the access must be evenly distributed among districts/cities in Indonesia.

Acknowledgements The authors gratefully acknowledge the Article 33 Indonesia research institute for the financial assistance and data used in this study. Article 33 Indonesia is a research institute engaged in the fields of education, the environment, and the development of vulnerable groups.

References

1. Bella A, Dartanto T (2018) Persons with disabilities (PWD) and poverty in Indonesia. Malays J Econ Stud 55(2):167–188. https://doi.org/10.22452/MJES.vol55no2.2
2. Byrnes A, Conte A, Gonnot J-P, Larsson L, Schindlmayr T, Shepherd N, … Zarraluqui A (2007) Disabilities—Handbook for Parlamentarians—from exclusion to equality: realizing the rigHts of persons with disabilities
3. Cameron L, Suarez DC (2017) Disability in Indonesia: what can we learn from the data? Monash University & Australian Government (August)
4. Getis A (1995) Cliff AD, Ord JK (1973) Spatial autocorrelation. London: Pion. Progress Hum Geogr 19(2):245–249. https://doi.org/10.1177/030913259501900205
5. Huo XN, Li H, Sun DF, Zhou LD, Li BG (2012) Combining geostatistics with moran's i analysis for mapping soil heavy metals in Beijing, China. Int J Environ Res Public Health 9(3):995–1017. https://doi.org/10.3390/ijerph9030995
6. Huripah E (2020) Towards an inclusive social welfare institution for disabilities: the case of Indonesia. Asian Soc Work J 5(1):18–28. https://doi.org/10.47405/aswj.v5i1.122
7. Kavanagh AM, Krnjacki L, Aitken Z, Lamontagne AD, Beer A, Baker E, Bentley R (2015) Intersections between disability, type of impairment, gender and socio-economic disadvantage in a nationally representative sample of 33,101 working-aged Australians. Disabil Health J 8(2):191–199. https://doi.org/10.1016/j.dhjo.2014.08.008
8. Kelejian HH, Prucha IR (2010) Specification and estimation of spatial autoregressive models with autoregressive and heteroskedastic disturbances. J Econ 157(1):53–67
9. Liu D (2018) How to create animated graphics to illustrate spatial spillover effects
10. Mont D, Cuong N (2013) Spatial variation in the disability-poverty correlation: evidence from Vietnam (48659)
11. Pokempner J, Roberts D (2001) Poverty, welfare reform, and the meaning of disability
12. Stock JH, Watson MW (2007) Introduction to econometrics

The Impact of Sky View Factor on Pedestrian Thermal Comfort in Tropical Context: A Case of Jakarta Sidewalk

Victorina Arif and Lin Yola

Abstract Rapid increase of high Urban Heat Island (UHI) intensity, as one of the main contributors to climate change, is an urgent environmental issue in high dense cities today. The development of outdoor environment influences the pattern and behavior of city dwellers. Climate and physical features parameters affect the thermal comfort of humans doing their outdoor activities, such as walking. However, climate factors in identifying the success of an outdoor design is not frequently discussed especially in the tropical context. This study aims to determine the effect of microclimate on the sidewalk in Sudirman and Thamrin Street with spatial variations on thermal comfort by using Envi-met and Rayman simulation to determine the effect of thermal comfort on the comfort of walking activities in outdoor space. The level of walking comfort was analyzed using the correlation model of Outdoor Thermal Comfort (OTC). This study pinpoints that Sky View Factor (SVF) significantly correlated to thermal comfort (T_{mrt} and PET). This study also reported that 5 pm is the most comfortable time for walking while the least comfortable is at 1 pm in less and moderate shaded areas.

Keywords Microclimate · Thermal comfort · Walking comfort · Jakarta sidewalk · Tropical context

1 Introduction

Increase of Urban Heat Island (UHI) intensity is currently an ongoing issue that occurs in high-density cities as one of the impacts of rapid urbanization. Urban development significantly modified the meteorological conditions of the surrounding area [1] and have an impact on the urban microclimate especially the rising of air temperatures [2]. There are two substantial scales to measure the urban heat islands, which are urban boundary layer (UBL) and the urban canopy layer

V. Arif (✉) · L. Yola
Department of Urban Development Studies, School of Strategic and Global Studies, Universitas Indonesia, Jakarta, Indonesia

L. Yola et al. (eds.), *Sustainable Architecture and Building Environment*, Lecture Notes in Civil Engineering 161,
https://doi.org/10.1007/978-981-16-2329-5_6

(UCL) [3]. UBL is the scale above the building roughness level, while UCL is the scale roughly between the ground and the roof level of the building known as a micro-scale. In this study, thermal comfort condition was considered as the UHI at the UCL scale.

Outdoor thermal comfort is predominantly and tacitly associated with the UHI phenomenon [4]. Thermal comfort is defined as level of human satisfaction based on a combined effect of the physical and climatic parameters [5]. Human thermal comfort sensation is influenced by four factors, namely air temperature, radiation temperature, humidity, and wind speed [6] as well as individual factors including clothing and activities [7]. In tropical climates, humans are more likely to look for places with cooler temperatures [8]. Poor thermal comfort in public spaces causes a low public interest in carrying out outdoor activities in public spaces, one of which is walking. Studies shown that there is a strong correlation between walkability and city sustainability. UHI that is not mitigated properly will cause a significant increase of household energy demand [9]. By this, walkability has been seen as the main basis for urban sustainability [10].

As the biggest metropolis in Indonesia, Jakarta continues to grow and develop rapidly especially in the central of Jakarta. UHI in Jakarta could be seen by the increasing of air temperature in 2014 up to 2–3 °C higher compared to 2001 [11]. The main issue on this study is the high-density environment and pedestrians in the central of Jakarta. Sudirman and Thamrin Street is the main road of the city center, and currently completed the government's master plan project to design a comprehensive and integrated transportation system, including the pedestrian networks. In order to evaluate the Sudirman and Thamrin sidewalk, this study aims to investigate the pedestrian in Sudirman and Thamrin sidewalk through spatial, microclimate and thermal comfort variables by using simulation. This study also analyses the effect of thermal comfort on the comfort of walking activities in outdoor spaces by using the Outdoor Thermal Comfort (OTC) correlation model.

2 Methods

This study was situated at Sudirman and Thamrin Street, the main roads of Jakarta city center. Jakarta has a tropical climate that is hot and humid. Along the Sudirman and Thamrin sidewalk, six points have been selected as case study. In this study, the location of the receptor points was chosen based on the function and character of the surrounding area which affects the activity of pedestrian and the spatial character such as closed, shaded and open areas to develop various possible conditions of thermal comfort.

This study uses two simulation models, namely Envi-met and RayMan. Envi-met simulation is used to calculate the microclimate data and value of mean radian temperature (T_{mrt}). Meanwhile, RayMan simulation is used to calculate the sky view factor (SVF) and calculating Physiologically Equivalent Temperature (PET). The simulation conducted on May, 1st 2020, during 12 h from 6 am to

6 pm. Configuration used in Envi-met model is a 10 × 10 m limitation and a 20 × 20 × 20 grid. Due to the limitations in the process of taking direct meteorological data during the COVID-19 pandemic in 2020, the microclimate data from Envi-met simulation are used in RayMan simulation. Subjects in the Rayman simulation used personal data as users and receptors (150 cm height, 50 kg body weight, and 25 years of age). The RayMan simulation is calculated at 3 different times: (1) morning at 7.00 WIB; (2) during the day at 13.00 WIB; (3) afternoon at 17.00 WIB. These times reflected typical times of morning and evening commutes, and time close to the midday when the radiation is the highest.

Statistical analysis was used to determine the effect of microclimate variables on thermal comfort on the Sudirman and Thamrin sidewalk. The correlation test is used to show a relationship between each variable. The variables analysed were T_{mrt}, Air Temperature (T_a), SVF, Wind Speed (v), and Relative Humidity (RH). Correlation analysis using Bivariate Pearson correlation with SPSS program. The data used were Envi-met model data which had previously been validated. Through this study, it would be known the impact of SVF, to microclimate and thermal comfort (T_{mrt} and PET) and how important of its impact on outdoor thermal comfort consequently.

3 Results and Discussion

In this analysis, we focused on the correlation of SVF with microclimate and thermal comfort based on simulation that have been performed using Envi-met and RayMan. Figure 1 shows the Sky View Factor (SVF) value calculated using the RayMan application to show conditions at the receptor point.

Based on Rayman simulations, point A and point F has the lowest SVF value of 0.136 and 0.393, therefore these two points are categorized as highly shaded area. As seen, point A has lowest exposure to the sky caused by building and vegetation. Meanwhile, Point C and D are categorized as moderate shaded area. Point B and point E are categorized as less shaded area. Comparing SVF maps of each points shown in Fig. 1.

Figure 2 shows the T_{mrt} distribution maps at 13.00 (maximum solar radiation). The colors indicate the mean radiant temperature within the area. The highest T_{mrt} is concentrated on the main road with T_{mrt} values is up to 15 °C higher than shadowing areas due to the direct solar radiation. The model in Point A and F perform better than other points. While, Point E has the highest T_{mrt} value on most of its surface. Table 1 shows the SVF value as environment variables and climatic variables in Sudirman and Thamrin Street.

The results indicate the correlation between the SVF value shown in Fig. 2. Distribution of mean radian temperature (T_{mrt}) at 13.00 in six receptor points where in those particular points have lower SVF value. This is rational because when there is no radiation, the heat released and reflected between surfaces in the

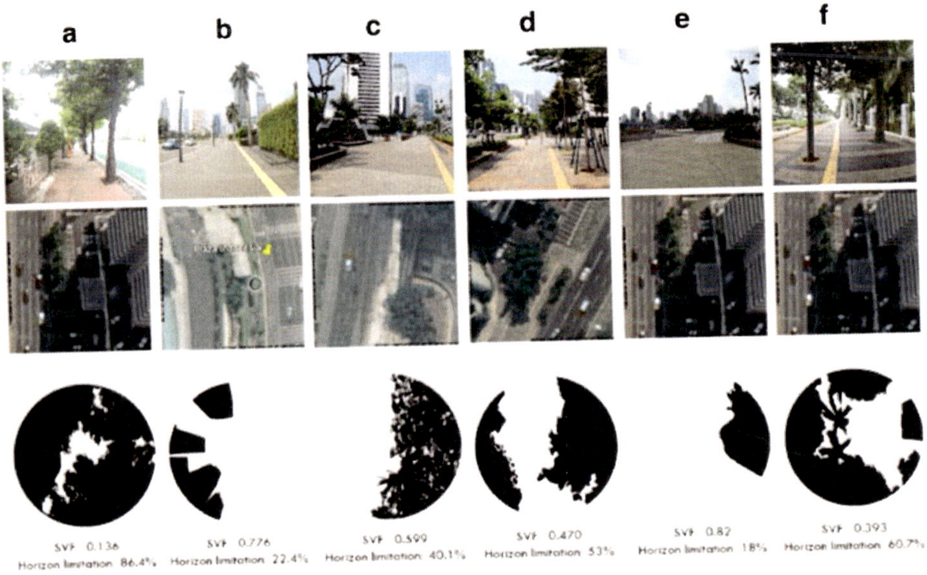

Fig. 1 Sky view factor calculations at 6 receptor locations using RayMan's simulations

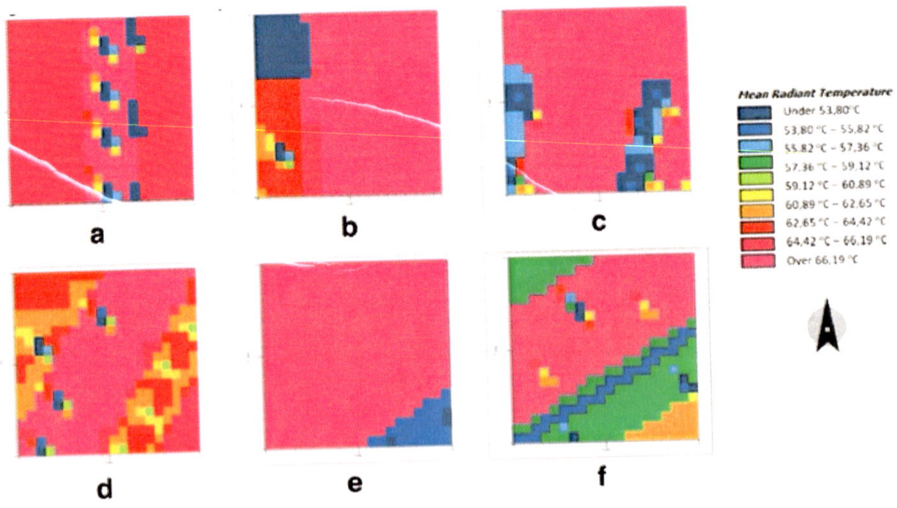

Fig. 2 Distribution of mean radian temperature (T_{mrt}) at 13.00 on six receptor points

form of long-wave radiation. To determine how far the sky view factor affects the thermal comfort, a correlation analysis is needed. The result is shown on Table 2.

Results quantitatively show that SVF is significantly correlated with SVF, while T_a, T_{mrt}, and PET with the value 0.905, 0.750, and 0.748. Based on the correlation results, SVF affects T_{mrt}, and PET in Jalan Sudirman and Thamrin. Thermal comfort will increase as the SVF value increasing. Correlation SVF and thermal

Table 1 Value of SVF, microclimate variables, and thermal comfort

Receptor points	SVF	T_a (°C)	RH (%)	v (m/s)	T_{mrt} (°C)	PET (°C)
A	0.136	25.23	94.55	2.14	55.05	23.50
B	0.776	25.38	93.32	2.07	58.75	29.87
C	0.599	25.36	94.38	2.01	58.77	30.00
D	0.470	25.34	94.43	1.93	55.98	29.93
E	0.820	25.49	93.13	1.94	58.83	30.10
F	0.393	25.24	96.05	1.87	58.84	30.20

Table 2 Pearson correlation coefficients between SVF and the microclimate and thermal comfort indices for all six points using Envi-met simulations

	SVF
T_a	0.905
RH	−0.689
v	−0.266
T_{mrt}	0.750
PET	0.748

comfort prove that the location in the less shaded area is much more get the sun penetration and will be hotter than the location in the shaded area. Different with negative value of correlation between SVF and relative humidity. This means that increasing of SVF will decreasing the humidity.

Using the model of Outdoor Thermal Comfort on walking (OTCw), the walking comfort value is estimated. The walking comfort is calculated at three different times and divided into three categories based on SVF value. Table 3 illustrates the results of walking comfort simulation at 7:00, 13:00, and 17:00. Based on simulation results, it is known that the highest discomfort on walking experience occur in moderate and less shaded area at 13.00 with the OTCw value is 5.1. Meanwhile, a more comfortable time on three locations achieved at 17.00 in the afternoon with OTCw value is 1.7 or the perception of warm.

In the walking comfort, it is indicated that the role of T_{mrt} and T_a are more dominant than the influence of wind speed. In this study it can be seen the

Table 3 The simulation results of walking comfort

Time	Highly shaded		Moderate shaded		Less shaded	
	OTCw	Preception	OTCw	Preception	OTCw	Preception
7:00	2.2	Hot/ uncomfortable	2.2	Hot/ uncomfortable	2.2	Hot/ uncomfortable
13:00	3.5	Very hot/very uncomfortable	5.1	Very very hot/ very very uncomfortable	5.1	Very very hot/ very very uncomfortable
17:00	1.7	Warm/slightly uncomfortable	1.7	Warm/slightly uncomfortable	1.7	Warm/slightly uncomfortable

significant effect of SVF on walking comfort. Receptor points with high shaded area have a better walking comfort level than the moderate and less shaded area. This shown at 13.00 when the solar radiation reaches its peak and causing OTCw value on moderate and shaded area increased up to 5.1 while the high shaded area only increased to 3.5. The lowest OTCw value occurs at 17.00 in three areas. This indicates that apart from SVF, the values of T_{mrt} and T_a are very influential on walking comfort level. Meanwhile, wind speed and humidity variables did not significantly affect walking comfort.

4 Conclusions

This research found that environmental parameters in the form of SVF have a significant correlation with thermal comfort. The result show that in Jakarta context, the exposure of area to solar radiation indicates higher air temperature (T_a), mean radiant temperature (T_{mrt}) and Physiologically Equivalent Temperature (PET). This also related to walking comfort that mostly effected by thermal comfort. Receptor points with high shaded areas have a better comfort level than moderate and less shaded areas, especially at 13.00 when the highest radiation occurs.

This research is based on the simulation data analysis and discussions. Using Envi-met and RayMan allow us to determine the environmental impact of the SVF. This also allowing us to have a better understanding on the experience of thermal comfort in existing city area. Further researches could include comparison between specified urban geometry changes, varied urban vegetation volumes, measurement on people experiences with different activities, and the impacts of different building surface coatings and ground surface materials. This analysis contributes to planning policies development with environmental strategies.

References

1. Landsberg HE (1981) The urban climate. Academic Press, Cambridge
2. Koerniawan MD, dan Gao W (2014) Thermal comfort and walkability in open spaces of mega kuningan superblock in Jakarta. In: The third international conference on countermeasures to urban heat Islands, IC2UHI, pp 654–666
3. Oke TR (2017) Urban climates. Cambridge University Press, Cambridge
4. Qaid A, Lamit HB, Ossen DR (2016) Urban heat island and thermal comfort conditions at micro-climate scale in a tropical planned city. J Energy Build 133(2016):577–595
5. Fanger PO (1970) Thermal comfort. McGraw-Hill, New York
6. Höppe PR (1999) The physiological equivalent temperature—a universal index for the biometeorological assessment of the thermal environment. Int J Biometeorol 24:699–706
7. ISO 7730 (2005) Ergonomics of the thermal environment—Analytical determination and interpretation of thermal comfort using calculation of the PMV and PPD indices and local thermal comfort criteria. International Standard

8. Zacharias J et al (2001) Microclimate and downtown open space. Environ Behav 33(2):296–315
9. Arifwidodo S, Chandrasiri O (2015) Urban heat island and household energy consumption in Bangkok, Thailand. Energy Procedia 79(November 2015):189–194
10. Southworth M (2005) Designing the walkable city. J Urban Plan Dev 131(4):246–257
11. Ramdhoni S, Rushayati SB, dan Prasetyo LB. (2016) Open green space development priority based on distribution of air temperature change in capital city of Indonesia, Jakarta. Procedia Environ Sci 33(2016):204–213

Pedestrian and Bicycle Lanes in Transit Oriented Development Area of Dukuh Atas, Jakarta

Dharmaraty Ashri Prawesthi and Lin Yola

Abstract The Indonesian authorities progressively promote Transit-Oriented Development (TOD) big cities, including DKI Jakarta. Specifically, The Governor of DKI Jakarta develops the TOD area in Greater Jakarta as urban regeneration of transportation. Dukuh Atas TOD was chosen as a pilot model as it has been determined through local regulations of DKI Jakarta in 2019. TOD concept aims to create an integrated area with pedestrian and cyclist accessibility. It basically accommodates the transit to the public transportation for a better quality of the environment; comfortable, safe, and lively. This study investigated principles of TOD; walking and cycling. Through a field survey, the indicators of walking and cycling were used to measure the success of the Dukuh Atas TOD. This study concluded that pedestrian and cycling lanes in the area, have not fully implemented. It concluded that the Dukuh Atas TOD has not yet integrated completely.

Keywords Pedestrian · Bicycle lane · Dukuh Atas · TOD

1 Introduction

TOD is the concept of developing areas within transit locations for the added value which centered on integration between mass public transport networks and non-motorized transportation mode networks. It is also aiming to reduce the use of motorized vehicles accompanied by the development of mix and dense areas with moderate to high space utilization intensity [1].

According to Calthorpe [2], the TOD concept is a concept that combines medium to high-density housing, with public functions, offices, trade, and services

D. Ashri Prawesthi (✉) · L. Yola
School of Strategic and Global Studies, Universitas Indonesia, Jakarta, Indonesia
e-mail: ashri.prawesti@ui.ac.id

L. Yola
e-mail: lin.yola@ui.ac.id

L. Yola et al. (eds.), *Sustainable Architecture and Building Environment*,
Lecture Notes in Civil Engineering 161,
https://doi.org/10.1007/978-981-16-2329-5_5

in a mixed-use development. Calthorpe sees these environmental characteristics as a guide to neo-traditional design for creating sustainable environments.

A TOD development is a mixed-use community development within walking distance of the transit node and central commercial area. TOD combines residential land use, trade, services, offices, open spaces, and public spaces in a walkable environment that makes it easier for people and users to travel by foot, bicycle, and other modes of public transportation [2].

The Institute for Transportation and Development Policy (ITDP), a Non-Government Organization (NGO), issues the basic principles of TOD. In the guideline entitled TOD Standard 2.1, there are 8 principles of TOD [3] that must be applied in the development of the TOD area, namely: (1) Walking; (2) Cycling/Cycle; (3) Connect/Connect; (4) Public Transportation/Transit; (5) Mix/Mix; (6) Compact/Densify; (7) Compact, and (8) Toggle/Shift.

TOD is being developed in Indonesia to solve urban transportation problems. The area can be rail-based, airport-based, or bus terminal based. With the proliferation of Transit Oriented Zones development and without neglecting the principles of TOD, especially walking and cycling, it is necessary to have research that focuses on the principles of walking and cycling which require the land in its development.

2 Methods

This research is qualitative research. The analysis was conducted descriptively by studying a determination of the Dukuh Atas TOD area associated with existing conditions or actual conditions in the area regarding pedestrian and cycling lanes. This study focuses on 2 (two) principles of 8 (eight) basic principles of TOD. The two principles are Walking and Cycling. Mainly, the two principle were chosen to reinforce the TOD concept as a compact area, which can be reached by walking and cycling within 5 (five) to 15 (fifteen) minutes.

The criteria to be discussed in this study are the parameters as presented in Tables 1 and 2.

The research area delineation is based on the determination of the TOD area by the DKI Jakarta Government and focuses on the primary activity centers of Dukuh Atas which are within a radius of ±500 m (calculated from the Dukuh Atas MRT station).

3 Discussion

The Dukuh Atas TOD area was considered to have had criteria such as: being designated as the center of activity; served by rail-based mass public transport and other urban public transport; located in an area with low disaster vulnerability accompanied by mitigation to reduce disaster risk. Here are some field findings.

Table 1 Aspects that affect pedestrians

Indicator	Variable	Parameter
Easy to access the Pedestrian lane	Availability of pedestrian lane	Pedestrian lanes that protect people from vehicles and other disturbances such as street vendors
		Directly available on the exit of the public transport station
Pedestrian lane connectivity	The area that is accessible by foot	100% available in all region
		Inter-connected between transit stations in the area
	Ideal walking distance	The maximum walking distance from and to transit stations and service centers is about <1 km
Friendly Pedestrian lane	Pedestrian lane dimensions	• The width of the main street is at least 3 m • Minimum width of 2 m on neighborhood roads (residential street/mixed use street)
	Secure Pedestrian lane	The material is unslippery and uses tactile for people with disability
		There are bollards for safety
		Adequate lighting
		There are road crossing facilities
	Convenience Pedestrian lane	There is shade (could be trees or an arcade and canopy)
		Shelters such as bus stops
		Rest areas including park benches/benches/another street furniture
		Food/drink stalls or kiosk at certain points
		Public toilet
		Public nursing room

Source The results of literature review that come from: Ditmass dan Ohland 2004 [4] and TOD Standards, 2013 [5]

Table 2 Aspects that affect cyclists

Indicator	Variable
Safety bike lane	For roads that have a driving speed of >30 km/h, the bicycle lane must be protected/separated from the direct vehicle lanes
	Roads that have a driving speed of <30 km/h do not need to use protected bicycle lanes, with markings are highly recommended
	Roads that prioritize pedestrians or shared streets with a speed limit of 15 km/h (pedestrian and bicycle lanes do not need to be separated)
	Lanes dedicated specifically to pedestrians and bicycle users

Source ITDP, 2017

3.1 Availability of Pedestrian and Bike Lanes

According to the Minister of Public Works Regulation No. 3/2014 concerning guidelines for planning, providing, and utilizing pedestrian network infrastructure and facilities in urban areas, explains that pedestrian lanes must be a connecting route between activity centers, block to block, and parcels to parcels in urban areas.

All blocks in the Dukuh Atas TOD area are previously connected to pedestrian lanes, especially on that direct connection to the highway. For neighborhood roads, some use the pedestrian lane, and some don't. Overall, based on survey results, the availability of pedestrian lanes in the area has only reached 60%. As for the bicycle lane, in the Dukuh Atas TOD area, there are only Sudirman Street, Tanjung Karang Street, and Teluk Betung Street (Fig. 1).

3.2 Pedestrian and Cyclist Lane Connectivity

From the observations, there is pedestrian access between the Trans Jakarta Bus Stop, the City Bus Stop, the KRL Sudirman Station, and the Dukuh Atas MRT Station. However, accessibility with BNI City Station is not yet connected (Fig. 2).

Based on the survey results, the travel time to Dukuh Atas MRT station on foot from blocks 3, 4, 6, 7, and 8 will average about 11.38 min. Meanwhile, for cycling from the same block, an average of 3.46 min. That may occur due to constraints on

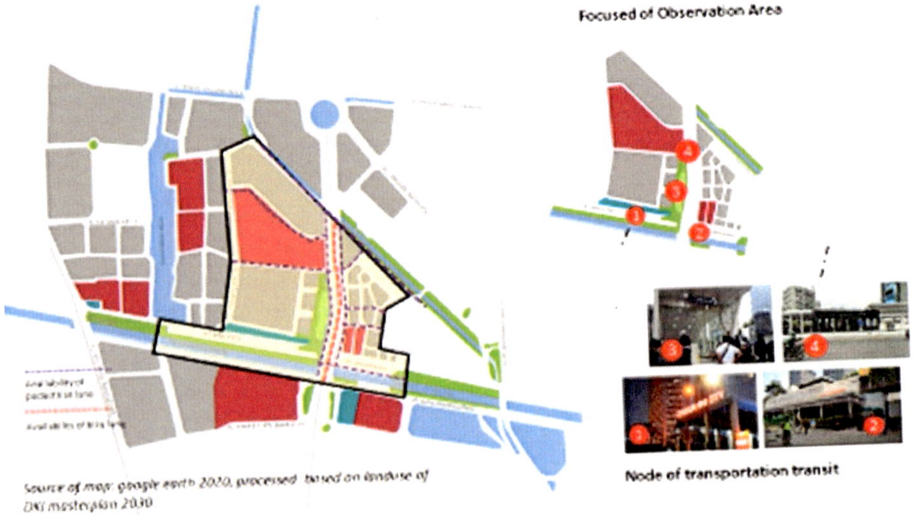

Fig. 1 Availability of pedestrian and bike lane in the Dukuh Atas TOD area. *Source* Map of RDTR DKI Jakarta, compiled

Fig. 2 Pedestrian line connectivity between Dukuh Atas MRT Station (**a** and **b**); Sudirman KRL Station (**c** and **d**); Trans Jakarta Bus Stop (**d**); unconnected pedestrian lanes at BNI City Station (**e**). *Image source* Documentation of field survey, 2019

the pedestrian lane that are not continuous or damage and become a place of street stalls.

Based on the survey results, the travel time to Dukuh Atas MRT station on foot from blocks 3, 4, 6, 7, and 8 will average about 11.38 min. Meanwhile, for cycling from the same block, an average of 3.46 min. That may occur due to constraints on the pedestrian lane that are not continuous or damage and become a place of street stalls (Fig. 3).

Fig. 3 Obstacles encountered in the movement of pedestrians and cyclists (Jalan Kendal and Jalan Blora). *Image source* Documentation of field survey, 2019

3.3 *Friendly Pedestrian Lane*

3.3.1 Pedestrian Lane Dimensions

The width of the pedestrian lanes still varies. The width found in the area is between 1.5 and 2 m (on neighborhood roads) to 4.5 m (on Jalan Sudirman) (Fig. 4).

3.3.2 Convinience

In general, the distribution of shade facilities is still lacking. Some lanes are still not shaded by shady trees or building canopies. It becomes heated during the day. Besides, there will be no place to take shelter during the rain.

Street furniture on the pedestrian lane, not yet fully available. Only the pedestrian lanes on the main road have street furniture such as benches, shelters, kiosks, and not equipped with public toilets.

Fig. 4 Dimensions of the Pedestrian Lane in Blocks 2, 3 and 4 of TOD Dukuh Atas. *Image source* Documentation of field survey, 2019

3.3.3 Security

Public Street Lighting (PJU) is already available on all pedestrian lanes in the area. Likewise, with the person crossing facilities in the form of crossing bridge and pelican crossing.

Security for people with disability is readily available but at best on large pedestrian lanes, which are close to transit points or roads.

3.4 Bike Lane Condition

The bike lane is not fully connected to the area. Only on the main road that previously has its own bike lane. In addition, there are 2 (two) bike stations, which are around the Dukuh Atas MRT Station and the Tosari Trans Jakarta Bus Stop (Fig. 5).

Based on the explanation above, the TOD Dukuh Atas area is still not optimal in applying the principles of walking and cycling. Most of the indicators used as references in this study are still not ideally applied to the Dukuh Atas transit areas.

The private building lots in the area are not open to the public and difficult to access for pedestrians. The provision of level crossing facilities, pedestrian bridges, and underpasses in the area is still limited, making it quite difficult for pedestrians who want to move from one block to another.

Fig. 5 Bike lane's condition in TOD Area of Dukuh Atas. *Image source* Documentation of field survey, 2019

4 Conclusions

The Dukuh Atas TOD area is an ongoing process of being developed into a better TOD area. Pedestrian and cycling lanes that should be applied to the area are still not running perfectly. Pedestrian and cycling lanes that should be provided to facilitate affordability in the area are not yet completely available.

Some concerns are needed to improve Dukuh Atas as a TOD area as follow:

1. It is necessary to conduct research with high technology to determine the movement of pedestrians and cyclists using GPS or Wi-Fi, and other sophisticated tools.
2. It is required to build a transit area infrastructure that can accommodate and allow pedestrian movement across the Dukuh Atas urban block which is separated by four sections by the canal and the Sudirman flyover road.
3. The development and additional length of the number of pedestrian lanes will be directed at all blocks in the Dukuh Atas transit area so that they can support walking activities in the transit area.
4. Eliminate the land guardrail found on each plot and create a dedicated passport for pedestrians to improve good connectivity.
5. Creating a safe and comfortable and friendly design for pedestrians and cyclists
6. Creating an inclusive design that is for all users.

The findings of this study were recommended for the DKI Jakarta authorities to improve the current Dukuh Atas TOD system and development. It was also suggested as a significant literature for the general planning of TOD area in DKI Jakarta or other big cities in Indonesia with the similar context.

References

1. Regulation of Minister of ATR/BPN No. 16 of 2017
2. Calthorpe P (1993) The next American metropolis: ecology, community, and the American dream. Princeton Architectural Press, Canada
3. Institute for Transportation & Develompent Policy (ITDP) Indonesia (2017) TOD standard volume 3.0. Accessed from http://www.itdp-indonesia.org/library/tod-standard-3/
4. Dittmar H, dan Ohland G (2004) The new transit town—best practices in transit oriented development. Island Press, Washington DC
5. Institute for Transportation & Develompent Policy (ITDP) (2013) TOD standard 2.1. New York

The Influence of Social Distancing Large Scale to Sustainability Micro Small and Medium Enterpries

Rosnaini Daga, Lin Yola, Bahtiar, and Andi Makkulawu Panyiwi Kessi

Abstract The Social Distancing Large Scale is a term of health quarantine in Indonesia which is defined as the limitation of certain activities of the population in an area that is suspected of being infected with a disease and/or contamination with possible forms to prevent the spread of disease or contamination. The aim of which is to prevent the spread of the Corona virus 19. The objectives of this study are (1) To find out how big the impact of the Large-scale Social Restriction policy on Micro, Small and Medium Enterprises in Makassar City, (2) To Know the Government's Handling of Micro, Small and Medium Enterprises due to the impact experienced by large-scale social distancing. The method used in this research is descriptive qualitative, according to Satori (Qualitative research method, alpabeth, 2011, [1]: 23) that qualitative research is carried out because the researcher wants to explore unquantifiable phenomena that are descriptive in nature such as the process of a work step, the formula for a recipe, the understanding of a variety of concepts, the characteristics of a good and service, an image, a style, a culture, physical model of an artifact and so on. Results research the implementation of large-scale social distancing in Makassar City, the trade sector, manufacturing industry, and the provision of accommodation, food and drink are at a negative growth level. Meanwhile, this sector is the most affected by the Covid-19 pandemic. Moreover, with the implementation of large Scale Social Distancing and the policy of working from home by a number of companies in Makassar City, the transportation sector has decreased by −51.15%, the provision of accommodation and food and drinks is

R. Daga (✉)
Institut of Bisnis and Finannce Nitro, Makassar, South Sulawesi, Indonesia
e-mail: rosnaini.daga@nitromks.ac.id

L. Yola
Department of Urban Studies, School of Strategic and Global Studies, Universitas Indonesia, Jakarta, Indonesia
e-mail: Lin.yola@ui.ac.id

Bahtiar
STIE AMKOP|School of Management and Business, Makassar, South Sulawesi, Indonesia

A. M. P. Kessi
STIE Tri Dharma Nusantara, Makassar, South Sulawesi, Indonesia

L. Yola et al. (eds.), *Sustainable Architecture and Building Environment*,
Lecture Notes in Civil Engineering 161,
https://doi.org/10.1007/978-981-16-2329-5_

−30.91%, company services are −27.34%, wholesale and retail trade by −8.29%, processing industry by −8.23%, and construction by −4.94%.

Keywords Social distancing · Large scale · Micro, small and medium enterprises

1 Introduction

The Social Distancing Large Scale is a term of health quarantine in Indonesian which is defined as "The restriction of certain activities of the population in an area suspected of being infected with a disease and/or contamination in such a way as to prevent the possibility of spreading disease or contamination. The Social Distancing Large Scale is a type of health quarantine implementation in the region, apart from home quarantine, hospital quarantine, and area quarantine". The Social distancing large scale objective is to prevent the spread of a public health emergency that is happening among people in a certain area. Restrictions on religious activities, and/ or restrictions on activities in public places or facilities. Social distancing large scale is carried out by local governments at the provincial and district/levels city after obtaining the approval of the Minister of Health through a Ministerial Decree.

The legal basis for regulation the Social Distancing Large Scale is Constitution Number 6 of 2018 concerning Health Quarantine. The law explains that further provisions regarding the criteria and implementation of the Social distancing large scale are regulated by a Government Regulation as a derivative regulation of the Law. To deal with the 2019 coronavirus disease which has become a pandemic, including in Indonesia, the government issued Government Regulation Number 21 of 2020 concerning The Social Distancing Large Scale in the Context of Accelerating Handling of Corona Virus Disease the year 2019 (Covid-19). In addition, the government also issued a Minister of Health Regulation Number 9 the year 2020 as a guideline for implementing The Social Distancing Large Scale. This Minister of Health regulations explains that Social Distancing Large Scale is implemented during the longest incubation period of Covid-19 (14 days) and can be extended if there is evidence of spread.

The origin of this pandemic came from Wuhan, China. Which one was found at the end of December 2019. And until now, according to World Health Organization (WHO) about data as of March 1, 2020, there have been 65 countries that have contracted this virus. The first According to epidemiological data, it shows that 66% of patients are related to a market or live market in Wuhan, China, of which patients show isolate samples that show a coronavirus infection, a new type of beta coronavirus, which is named the novel Coronavirus-19.

2 Literature Review

2.1 *Large Scale Social Distancing*

The Large-Scale Social Distancing is a policy issued by the central government to reduce or minimize the spread of the Covid-19 virus that is currently hitting Indonesia and even the world. If seen technically, it can determine that The Large-Scale Social Distancing is a certain action against a community or area where the area or area is suspected of being contaminated by a disease, and is intended to prevent the spread of a disease. Based on the warrant from the Ministry of Health 2020, it is said that the Large-Scale Social Distancing is a type of regional health quarantine, apart from including home, regional and hospital quarantine.

Based on the Government Regulation of the Republic of Indonesia No. 21 of 2020 concerning The Large-Scale Social Distancing, the objective is to prevent the spread of the corona virus from spreading, and also it is included in the category of public health emergencies.

The implementation of the Large-Scale Social Distancing policy is expected to be able to stop the growth and spread of Covid-19 in the regions, and it is hoped that this corona virus can be eliminated quickly. Not only that, the policy also serves the following matters including:

(a) Maintaining public health, where the current corona outbreak can be stopped is implementing social distancing
(b) Social safety nets, where the government provides assistance to meet community needs during the corona outbreak, and also to maintain people's purchasing power
(c) Maintain the business world, provide assistance to MSMEs to keep doing their business and look after their workforce
(d) Assistance from the lower layers of society, which are also the people most affected by covid-19 are the lower classes, therefore the government provides assistance

2.2 *Micro, Small and Medium Enterprises*

Micro, Small and Medium Enterprises is one of the main pillars a national economy with an independent perspective has great potential for improve well-being. Micro small and Medium Enterprises (MSMEs) is one of the main pillars insightful national economy independence has great potential for improve well-being. Role Micro, Small and Medium Enterprises, especially since the monetary crisis last year 1997 can be viewed as media a rescuer in the recovery national economy. In supporting Micro, Small and Medium Enterprises activities, the government does support through Facebook policies expected to give injection to Micro, Small and Medium Enterprises o as not to stop in the middle of the road.

Yola and Siong [2], Results the research present that the airflow found very weak in four urban configuration. However, the setting of vertical obstruction in the urban configuration helps to improve the ventilation. The study recommends that the canyon feature develops a better airflow in urban configuration.

Micro business is classified as a type of business marginal, which is due to the use of technology relatively simple, capital level ones low, low access to credit, and tends to be market-oriented local. Therefore, it always has to be strived for the right strategy for empowering MSMEs so that people's welfare is increasingly elevated. Various the role strategic of the MSMEs sector, but this sector is also faced with various the problem. In its application found related problems which experienced by MSMEs, among others (1) Capital money, (2) Difficulties in marketing, (3) intense business competition, (4) Difficulty in raw materials, (5) less technical production skills, (6) lack of skills managerial efforts and quality of resources human management, (7) less financial management, (8) business climate (licensing, laws and regulations) which is less conducive. Constraints and small and informal business issues others are also due to difficulty access to information and resources productive like capital and technology, that is resulting in limited ability small business to develop [3].

3 Method

The method used in this research is descriptive qualitative, according to Satori ([1]: 23) that qualitative research is carried out because the researcher wants to explore unquantifiable phenomena that are descriptive in nature such as the process of a work step, the formula for a recipe, the understanding of a variety of concepts, the characteristics of a good and service, an image, a style, a culture, physical model of an artifact and so on. In this study, two sources of secondary data collection were used. Secondary data is a data source that does not directly provide data to data collectors, but sees other people or with document.

Secondary data is data that comes and is loaded from information that has been collected from an existing source. Secondary data sources can be found or come from corporate or company records or documentation, government publications, industry analysis by media, websites, online-based documents, and so on.

The collection of secondary data in this study was carried out by research and recording documents, among others, by looking for verified journals and published on websites that have guaranteed authenticity, as well as in several textbooks, as well as government regulations.

4 Result and Research Description

Impact of the Covid-19 Pandemic on the Business Sector as well Implications for Economic Growth and Several Indicators Development in Makassar City. The trade and transportation sector experienced a significant decline. There are 3 (tree) main challenges faced by MSME actors during the Covid-19 pandemic that has plagued the country. "First is in terms of supply. Since the implementation of Large-Scale Social Restrictions in almost all regions, many of our MSMEs have had difficulty getting raw materials".

The second challenge is the cash flow problem. Many MSME players feel that their income has decreased due to the absence of customers who have purchased products since the PSBB and physical distancing were implemented, and The third challenge is from falling demand. Market uncertainty makes the demand for goods sold by MSMEs influential.

The onslaught of the Corona Virus Disease or Covid-19 outbreak around the world also has an impact on the Micro, Small and Medium Enterprises (MSMEs) sector in South Sulawesi, especially in Makassar City. MSMEs are experiencing various dire conditions, some are temporarily closed, some are still open but customers are very minimal, the point is that MSMEs' income during the Pandemic period, especially during the large-scale social distancing period, dropped dramatically, even some MSMEs were threatened with bankruptcy.

The pandemic period is due to the corona virus or commonly known as Covid-19. Several preventive efforts have been made, starting from work from home (WFH) and always using a hand sanitizer and a mask when traveling. Of course, this does not have a significant effect on reducing the spread of this virus, therefore, on April 1, 2020, President Indonesian made a policy, namely Social Distancing Large Scale or known as Large-Scale Social Restrictions. With the hope that this virus will quickly reduce it spread.

Beside, this policy implement restriction which include dismissing school children, dismissing some workers, and also limiting access to public transportation and also public facilities and also this policy prohibits us from gathering and always wearing masks.

From this policy also emerged several effects which hindered economic growth. There were several price decreases, such as the price of fish, vegetables and other food prices, due to decreased demand. For example, a restaurant business that demands goods is not running as usual. This is intended to make this policy effective in reducing the spread of the corona virus which has caused many losses.

With the first phase of Social Distancing Large Scale implementation in Makassar City, all MSMEs are required to close, this makes MSMEs not operating and there is no income during the Social Distancing Large period. So that it has an impact on MSMEs in South Sulawesi, it is recorded that there are 3022 Micro, Small and Medium Enterprises (MSMEs) affected. This figure is based on data compiled from the Provincial Government in this case the South Sulawesi Cooperative and, Small and Medium Enterprises (MSMEs) Office as of May 1,

2020. Thousands of these businesses had to go out of business during the pandemic. It is different from the implementation of Social Distancing Large Scale phase II, where the Makassar City Government asks Micro, Small and Medium Enterprises to remain open to the implementation of Large-Scale Social Restrictions in Makassar City which also triggers people not to visit shopping malls as a form of anticipation. Meanwhile, the decline in the number of visits to shopping centers occurred in the city of Makassar, in September to 58%, down from 66% in August. The impact of the Covid-19 or Corona Virus pandemic is felt in all sectors of the economy.

There are 6 (six) business sectors that are most affected by the outbreak of COVID-19 in Makassar City. Three of them, namely the trade, manufacturing and construction sectors, are the main pillars of Makassar's economy, with an average contribution of more than 10% per year in the last decade. Meanwhile, the other 3 (three) affected sectors are the accommodation, food and drink provision sector, the corporate services sector and the transportation sector. Based on the results of the implementation of the Social Distancing Large in Makassar City, the trade sector, manufacturing industry, and the provision of accommodation, food and drink are at a negative growth level. Meanwhile, this sector is the most affected by the Covid-19 pandemic. Moreover, with the implementation of Social Distancing Large and the policy of working from home or WFH by a number of companies in Makassar City, the transportation sector has decreased by –51.15%, the provision of accommodation and food and drink by –30.91%, company services of –27.34%, wholesale and retail trade by –8.29%, processing industry by –8.23%, and construction by –4.94%.

Currently, Makassar City's economy has started to enter the recovery phase. Although the Covid-19 pandemic is not over, recovery efforts have begun to be driven by the Makassar City Government, which is currently starting to focus on improvements in five sectors. The improvement in the five sectors follows a number of easing made by the government to revive economic activity. The Makassar City Government itself has made Regulation of Mayor Number 51/2020 concerning the Implementation of Discipline and Law Enforcement of the Health Protocol as an effort to prevent Covid-19 in the Framework of Economic Recovery and Acceleration.

Makassar's economy, which had grown at 6.20% at the beginning of the year, had to fall to 2.8% in the second quarter of 2020. Makassar's economic growth in the third quarter was only 3%. Although it is still slowing down, there have been improvements from the previous period.

This is because micro, small and medium enterprises (MSMEs) have been heavily affected by the Covid-19 pandemic, especially with the implementation of Social Distancing Large. From the start of the implementation of the Social Distancing Large scale until it was completed at the second Social Distancing Large, small and medium enterprises (MSMEs) also needed direct cash assistance. The Office of Cooperatives and SMEs has provided assistance to 8584 MSMEs, the assistance in the form of venture capital assistance of IDR 2.4 million to each affected MSMEs.

During the Pandemic period, many MSME players switched businesses to become craftsmen of masks, hand sanitizer, culinary, making herbal concoctions that can increase the body's immunity. The business results of these MSME players are sold online by using online transportation or goods delivery services because the community is still very limited in making direct transactions. Moreover, the Enforcement of Social Distancing Large in Makassar City is approaching the month of Ramadan, which in previous years public consumption of products sold by MSMEs has increased, but in 2020 during the Pandemic period with the enactment of the Enforcement of Social Distancing Large Public consumption in the month of Ramadan is not as usual, it has even decreased dramatically because people are afraid to leave the house, let alone make food shopping transactions, people are afraid of being exposed to the Covid-19 virus. This causes the turnover of MSMEs players to also drop dramatically.

The imposition of Social Distancing Large in Makassar City had a negative impact on the growth of MSMEs in Makassar City, South Sulawesi. There are 6 (six) business sectors that are most affected by the outbreak of Covid-19 in Makassar City. Three of them, namely the trade, manufacturing and construction sectors, are the main pillars of Makassar's economy, with an average contribution of more than 10% per year in the last decade. Meanwhile, the other 3 (three) affected sectors are the accommodation, food and drink provision sector, the corporate services sector and the transportation sector. Based on the results of the implementation of the Social Distancing Large in Makassar City, the trade sector, manufacturing industry, and the provision of accommodation, food and drink are at a negative growth level. Meanwhile, this sector is the most affected by the Covid-19 pandemic. Moreover, with the implementation of Social Distancing Large and the policy of work from home by a number of companies in Makassar City, the transportation sector has decreased by –51.15%, the provision of accommodation and food and drink by –30.91%, company services of –27.34%, wholesale and retail trade by –8.29%, processing industry by –8.23%, and construction by –4.94%.

By considering the capabilities of the resources owned and the level the urgency of the COVID-19 pandemic, the Makassar City Government is expected to be soon propose the status of Large-Scale Social Restrictions in Makassar. That effort so far done by the Makassar City Government, in this case Social Restrictions Small Scale, does not have a clear juridical foundation. Apart from that, generally run like a regional quarantine concept.

However, to streamline the implementation of the Large-scale Social distancing in Makassar, the government it is necessary to pay attention to three things, namely: (1) The government needs to make appropriate plans adequate related to the handling of COVID-19 health within 14 days; (2) Good coordination with legal officials so that the large scale Social distancing in Makassar can be implemented firmly in accordance with the applicable rules; (3) Ensuring the availability of MSMEs needs and divided effectively and evenly.

It can be said that this large scale social restriction policy has a great impact on micro, small and medium enterprises. It can be seen once that there is a significant drop in turnover, even many micro, small and medium enterprises were forced to close during the implementation of large-scale social restrictions in Makassar, which of course greatly affects the survival of micro, small and medium enterprises.

5 Conclusions

The imposition of Social Distancing Large in Makassar City had a negative impact on the growth of MSMEs in Makassar City, South Sulawesi. There are 6 (six) business sectors that are most affected by the outbreak of Covid-19 in Makassar City. Three of them, namely the trade, manufacturing and construction sectors, are the main pillars of Makassar's economy, with an average contribution of more than 10% per year in the last decade. Meanwhile, the other 3 (three) affected sectors are the accommodation, food and drink provision sector, the corporate services sector and the transportation sector. Based on the results of the implementation of the Social Distancing Large in Makassar City, the trade sector, manufacturing industry, and the provision of accommodation, food and drink are at a negative growth level. Meanwhile, this sector is the most affected by the Covid-19 pandemic. Moreover, with the implementation of Social Distancing Large and the policy of working from home (WFH) by a number of companies in Makassar City, the transportation sector has decreased by –51.15%, the provision of accommodation and food and drink by –30.91%, company services of –27.34%, wholesale and retail trade by –8.29%, processing industry by –8.23%, and construction by –4.94%.

References

1. Satori AKD (2011) Qualitative research method, alpabeth. Bandung, Indonesia
2. Yola L, Siong HC (2018) Impact of urban canyon direction on solar radiation and airflow in hot and humid. Asian J Behav Stud 3(3). http://dx.doi.org/10.21834/ajbes.v3i13.146
3. Kurniawan FD, dan Fauziah (2014) Empowerment of micro, small and medium enterprises (MSMEs) in poverty alleviation. JKMP 2(2)

Empirical and Numerical Approaches in Urban Microclimate Modeling: Investigation on the Reliability

Lin Yola, Olutobi Gbenga Ayegbusi, Halvina Grasela Saiya, and Komara Djaja

Abstract Urban climate modeling indicates a significant development in accomplishing the urgent solutions to the current urban climate change challenge. As the demand for sustainable urban development continue to grow, researchers have derived range of strategies and methodologies to explore the urban issues. Direct field measurements and data collection have been the popular approach for urban microclimate studies. However, in recent times, numerical simulation approaches have become increasingly popular. This study presents the urban microclimate modeling of a tropical outdoor space with dense building blocks using empirical and numerical approaches. This investigation covers the evaluation of air temperature, air movement, and relative humidity. The results revealed that the two approaches are significantly correlated. However, this study points out that numerical modeling is a reliable approach in performing the hypothetical urban microclimate modeling in the high-rise building blocks. The findings therefore suggest that numerical approach is better in the study of Urban Heat Island for strategic mitigation purpose than empirical approach.

Keywords Urban configurations · Urban microclimate · Reliability study · ENVI-met simulation · Tropical context

L. Yola (✉) · K. Djaja
Department of Urban Studies, School of Strategic and Global Studies, Universitas Indonesia, Jakarta, Indonesia
e-mail: lin.yola@ui.ac.id

O. G. Ayegbusi
Department of Architecture, Faculty of Built Environment and Survey, Universiti Technologi Malaysia, Jakarta, Indonesia

H. G. Saiya
School of Environmental Science, Universitas Indonesia, Jakarta, Indonesia

L. Yola et al. (eds.), *Sustainable Architecture and Building Environment*, Lecture Notes in Civil Engineering 161,
https://doi.org/10.1007/978-981-16-2329-5_7

1 Introduction

The effect of climate change is becoming more prevalent as the city grows and make necessary provision for the increasing population. Urban Heat Island (UHI) appears to be an ongoing issue in big cities since the 1818 to 1925 as the result of continuous rapid economic activities in the urban area and is one of the major contributors to climate change [1–4]. Apart from the environmental degradation caused by UHI, poor thermal comfort, urban health problem, and high cooling loads are among other severe impact of this phenomenon which require closer attention and mitigation efforts from both professional planners and policy makers to achieve a sustainable urban environment [3–9].

The mitigation effort widely opens the opportunities for scholars to investigate the urban climate to map the alternatives of strategies and actions. This study justifies that microclimate modification is an essential component to form the UHI, besides it is also a uniform measure in assessing the urban energy balance [3, 10]. The micro-urban spaces between buildings generate urban heat generators through short and longwave radiation, surface heat storage, heat canopy, and anthropogenic heat [10]. As the most used variable in urban energy balance study, microclimate significantly affects indoor and outdoor thermal comfort. In general, modification of microclimate mainly focuses on the following variables; solar radiation, air temperature, air velocity, and relative humidity [11]. The main objective of the urban climate study is to minimize the microclimate modification that increases the UHI intensity, which would also reduce the thermal discomfort and demands of cooling loads. Therefore, the methodology of conditioning the microclimate modification is constantly explored. As each city is attached to its context in adapting to the regional climate, the approach frequently varies between one to another. This study however looks closer into the analysis of the most reliable approach to investigate the urban microclimate modeling of a dense tropical city through a comparison study.

2 Urban Microclimate Modeling

As the study on urban climate develops, Oke [12] as one of the pioneers in this area stressed three major methods in the UHI investigation; observation, theoretical, and modeling. As the focus of this study, the modeling method described as a 'simplified presentation of reality to the current theory [13]. Atkinson [14] grouped UHI modeling into the following approaches; hardware, physical equation, and dynamical numerical. This study reviewed that the category outlined by Oke [15], Svensson et al. [16], and Voogt [7] is the most discussed and relevant in the UHI modeling are as follows; numerical, physical, and empirical.

The empirical modeling method is commonly used in early urban climate study. As the main focus, the UHI investigation was performed through conventional field observation. Empirical modeling was formed by the statistical field data that then formulated into mathematical equations [17, 18]. A significant study using empirical modeling in investigating the relationship between urban canyon geometry and UHI was conducted by Oke in 1981 [19]. From this study, Oke formulated that maximum intensity of UHI in canyon space is resulted by the geometry of high Height to Width (H/W) ratio and low Sky View Factor (SVF) value [11, 20], as shown in Eqs. 1 and 2. The formulas are used as the basic model in this study; particularly in closely investigate the urban temperature behavior in relation with the UHI scenario.

$$\mathrm{dTmax} = 7.45 + 3.97 * \ln(\mathrm{H/W}) \tag{1}$$

$$\mathrm{dTmax} = 15.27 - 13.88 * \mathrm{SVF} \tag{2}$$

Physical modeling is an approach by using a scale model to represent the real scenario of urban features. Oke [19] was also among the ones to extends the empirical modeling to physical modeling. Numerical modeling applies the mathematical model into a computer simulation. The rapid development of computer technology provides solutions to the constrain of the both empirical and physical modeling approach. It simulates a more accurate and precise physical and climate features, as well as it eliminates the limitation of time, cost, and research variables Batty [21] strongly described that the computer modeling as comprehensive, detail, and allow planners to perform the more flexible hypothetical urban climate model. Bruse [22] was among the scholars that pioneered the urban climate study by developing the computer model.

The computer simulation as the form of numerical modeling is the recent approach mostly applied to study the complex and dense urban climate. Besides, computer simulation covers most of the urban climate modeling limitations. The computer simulation does not only analyze the existing urban climate scenario but also a hypothetical scenario that eliminates time challenges. Therefore, the computer simulation is a trend approach among planners and scholars to assess the relationship between urban physical features and climate variables to achieve the most effective planning decision [23–26].

Yet, there is no one particular software covers all of the microclimate or thermal comfort variables [27]. The rapid development and intense use of computer simulation in nowadays urban climate studies do not stop the effort of scholars to include investigation on its' reliability to perform the current urban scenario with more complexity and meeting the contextual research objectives. The study of reliability or compatibility of the urban simulation tends to always perform to test the accuracy, effectiveness, and prediction of errors of the modeling approach. In short, the reliability of the numerical modeling is often performed by doing validation investigation with other modeling approaches. The example of the reliability study could be seen by comparing it with other simulation and field observations. In the context of this study, the reliability investigation seeks to closely analyze the gap of computer simulation results with the field data.

3 Methods

This study is an extended discussion of the previous study [28–31] that validates ENVI-met simulation through Oke's model. The previous study aimed to confirm the reliability of the ENVI-met from the early empirical mathematical model. Thus, this study aims to elaborate on the reliability of the ENVI-met simulation by comparing it with the field observation data. This study investigates two sites of high rise residential urban blocks, situated in a dense city center of Kuala Lumpur, Malaysia. The first site is a 19 storey block (Flat Bandar Tasik Selatan) while the second site is a 15 storey block (Surya Magna). The building configuration in the two sites is a Courtyard Canyon, with the canyon feature in the middle of the courtyard. However, the direction of the canyon in Flat Bandar Tasik Selatan faces East-West while in Surya Magna faces North-South.

The field data observation was conducted on 15 June 2015, set in the middle of the courtyard. The 24-hours data was taken by using the VelociCalc Plus 10 cm from the surface. This study presents data analysis of one of the three microclimate variables; air temperature. With the existing scenario, this study investigated four urban configurations; Courtyard, U configuration, Courtyard Canyon, and Canyon. The Sky View Factor (SVF) varies in each configuration while the Height to Width (H/W) aspect ratio remains constant. The simulation used ENVI-met V.3. The climate data was inserted in the ENVI-met simulation configuration editor are as follow: the initial temperature and inside temperature are 303.15 K and 293 K; relative humidity is 83%; albedo of walls and roofs are 0.3 and 0.5; the wind speed and direction are 1.4 m/s and 225 (South West to North East), and heat transmission wall and roofs are 1.94 W/m^2 K and 6 W/m^2 K). This study used the 250 × 250 × 30 size of simulation sections as the output data.

4 Results Discussion

This urban configuration and microclimate investigation presented the validation study [31]. It justified the air temperature behavior within the canyon space from the UHI model. It compared Oke's Model (Eqs. 1 and 2) and ENVI-met simulation. The results (Tables 1 and 2) pointed out that the drop of diurnal to nocturnal air temperature developed as the SVF value increases. This situation was as presented by dTmax in Oke's model, the UHI intensities were lower when the SVF value increases. Through comparison with the Oke's model as the empirical modeling, the results validated that Envi-met urban simulation was a suitable approach for this study.

The further reliability study on the simulation approach is presented in this paper. The results of this reliability study are also discussed according to the microclimate data from each case site. Overall, regression analysis on the two case sites of Flat Bandar Tasik Selatan (FBTS) and Surya Magna (SM) shows the

Table 1 Comparison of Oke's model and Envi-met: East–West

Urban configuration and SVF	Oke's model $dT_{max} = 15.27 - 13.88 * SVF$ (°C)		Drop of Envi-met air temperature (diurnal–nocturnal) (°C)	
Courtyard (0.275)	11.45	dT_{max} decreases when SVF increases	0.27	Drop of air temperature develops by the increase of SVF [a]U configuration shows higher air temperature when faces West
U (0.309)	10.98		0.78[a]	
Courtyard Canyon (0.438)	9.19		0.45	
Canyon (0.676)	5.89		1.32	

Source Yola et al. [31]

Table 2 Comparison of Oke's model and Envi-met: North–South

Urban configuration and SVF	Oke's model $dT_{max} = 15.27 - 13.88 * SVF$ (°C)		Drop of Envi-met air temperature (diurnal–nocturnal) (°C)	
Courtyard (0.611)	6.79	dTmax decreases for the higher SVF	1.13	Drop of air temperature increases as the SVF develops
U (0.694)	5.64		1.19	
Courtyard Canyon (0.707)	5.46		2.5	
Canyon (0.793)	4.26		2.96	

Source Yola et al. [31]

significant correlations between the two discussed empirical and numerical approaches (Table 3). However, the analysis indicates the small gap between the simulated and observed microclimate data. This situation is due to the unrecorded variables in the existing scenario that was not recorded in the simulation. Among the variables are building material properties, anthropogenic heat from vehicles and air conditioning, green area, and other vertical obstructions in sites.

The significant correlation findings from regression analysis point out that the computer simulation and field observation are both valid approaches to study the

Table 3 Microclimate regression analysis of observation and simulation

Canyon direction	Reg. analysis	Air temperature	Relative humidity	Wind velocity
East–West	P value	0.000	0.000	0.000
	R^2	0.709	0.834	0.939
	RMSE	0.140 °C (0.4%)	3.460% (4.6%)	0.035 m/s (4.3%)
North–South	P value	0.000	0.002	0.022
	R^2	0.977	0.804	0.602
	RMSE	0.262 °C (0.7%)	1.60% (2.2%)	0.015 m/s (2.2%)

microclimate study. Specifically, this study stresses that the computer simulation presents the identical pattern of result with the field observation with the tolerable margin error. In the context of this study, the field observation data validates the ENVI-met simulation as an effective tool of the numerical approach to investigate the microclimate in the dense urban configurations.

5 Conclusions

The development of urban climate study is a part of the positive response to numerous effects of climate change which requires urgent attention from both city planner and policy makers. Thus, understanding valid approach(es) to reliably explore the implications of urban growth and the resulting challenges is an essential stage in urban climate study development. This will provide a range of opportunities to generate comprehensive strategies to address climate change mitigation actions. The regression analysis was adopted in this study to investigate the correlation between the empirical and numerical approach, by comparing the results of the observed and simulated microclimate data of a hypothetical urban configuration in dense tropical residential site contexts. The results showed a significant correlation between the two approaches. The finding shows that ENVI-met simulation provides a strongly reliable results in the assessment of urban microclimate variables. It can also be inferred for the present study that the reliability of these results is due to ENVI-met efficiency, the available features, and the level of accuracy the technology brought to bear in modelling both two and three-dimension urban climate data. Finally, ENVI-met offers the opportunity to model a hypothetical urban configuration which is a limitation of other empirical modelling methods. Therefore, this study provides a scientific justification and the reliability of the results and hence strongly recommend the ENVI-met as an alternative approach in urban climate study.

References

1. Sachindra DA, Ng AWM, Muthukumaran S, Perera BJC (2015) Impact of climate change on urban heat Island effect and extreme temperatures: a case study. Q J Royal Meteorol Soc
2. Fatwa N (2020) Strengthening the role of Sharia Public banking in the Indonesian construction industry: towards an atmosphere of sustainable urban development. IOP Conf Series Earth Environ Sci 436(1):
3. Gartland LM (2008) Heat islands: understanding and mitigating heat in urban areas. Earthscan
4. Schmauss A (1925) Eine miniaturpolarfront. Meteorol Z 42:196
5. Arifwidodo S, Chandrasiri O (2015) Urban heat island and household energy consumption in Bangkok, Thailand. Energy Proc 79:189–194
6. The Heat Island Group (2005) Retrieved February 10, 2015, retrieved from http://eetd/lbl/gov/HeatIsland/
7. Voogt J (2004) Urban heat island: hotter cities. American Institutes of Biological Sciences

8. Santamouris M, Papanikolaou N, Livada I, Koronakis I, Georgakis C, Argiriou A, Assimakopoulos DN (2001) The impact of urban climate on the energy consumption of buildings. Sol Energy 70(3):201–216
9. Oke TR (1987) Boundary layer climates. Routledge, New York
10. Weber F (2015) Towards mitigative buildings and urban environments. In: The 31th international PLEA conference passive low energy architecture, 9–11 September 2015. Bologna
11. Erell E, Pearlmutter D, Wiliamson T (2011) Urban microclimate: designing the spaces between buildings. Earthscan
12. Oke TR (1982) The energy basic of urban heat island. J Royal Metereol Soc 108(455):1–24
13. Liu Y (2009) Modelling urban development with geographical information systems and cellular automata. CRC Press, Boca Raton
14. Atkinson BW (2003) Numerical modelling of urban heat-island intensity. Bound Layer Meteorol 109(3):285–310
15. Oke TR (1984) Towards a prescription for the greater use of climatic principles in settlement planning. Energy Build 7(1):1–10
16. Svensson M, Eliasson I, Holmer B (2002) A GIS based empirical model to simulate air temperature variations in the Göteborg Urban Area during the night. Climate Res 22(3):215–226
17. Masson V (2006) Urban surface modeling and meso-scale impact of cities. Theoret Appl Climatol 84:35–45
18. Balázs B, Unger J, Gál T, Sümeghy Z, Geiger J, Szegedi S (2009) Simulation of the mean urban heat island using 2D surface parameters: empirical modeling, verification and extension. Meteorol Appl 16(3):275–287
19. Oke TR (1981) Canyon geometry and the nocturnal urban heat island: comparison of scale model and field observations. J Climatol 1:237–254
20. Givoni B (1998) Climate considerations in buildings and urban design. Van Nostrand Reinhold
21. Batty M (2013) The new science of cities. Massachusetts Institute of Technology
22. Bruse M, Team (2015) ENVI-met website. Available at http://www.envi-met.com/. Retrieved on 2nd Feb 2015
23. Yola L, Siong HC (2014) Modelling and simulation approach of assessing urban form. Emerging trends in commerce, economics, humanities and social sciences (CEHSS). Int Res J Humanit Environ Issues, 25–26 Feb 2014. Kuala Lumpur
24. Yola L, Siong HC (2015) Validation of 3D microclimate simulation as quantitative analysis for urban configuration study in hot and humid region. International joint conference. SENVAR-iNTA-AVAN, 22nd–27th July 2015. Universiti Teknologi Malaysia
25. Yola L, Siong HC (2016) Computer simulation as an alternative approach in climatically responsive urban configuration study. International conference of low carbon Asia (ICLCA) 2016, 23rd–25th Nov 2016. Universiti Teknologi Malaysia
26. Yola L, Siong CS (2017) Computer Simulation as an Alternative Approach in Climatically responsive urban configuration study. Chem Eng Trans 56(2017):505–510
27. Vidmar J, Roset J (2013) Evaluation of simulation tools for assessment of urban form based on physical performance. Environmental physics for architects. Faculty of Architecture, Ljubljana, Slovenia
28. Yola L, Siong HC (2017) Computer simulation as an alternative approach in climatically responsive urban configuration study. Chem Eng Trans 56:505–51
29. Yola L, Siong HC, Djaja K (2020) Climatically responsive urban configuration in residential area: research gaps. AIP Conf Proc 2020(2255):0700141
30. Yola L (2020) Canyon effects in urban configurations: tropical context study. IOP Conf Series Earth Environ Sci 436(1):012028
31. Yola L (2018) Impact of urban configurations on microclimate and thermal comfort in residential area of Kuala Lumpur. PhD dissertation. Universiti Teknologi Malaysia

Evaluation of Urban Resilience Performance Indicator Frameworks in Indonesia

Raditya Pranadi, Triarko Nurlambang, and Lin Yola

Abstract Cities in Indonesia have continued to develop into centers of activity and their position in disaster-prone areas such as fault zones and coastlines leads to higher risks to various threats of natural and social disasters that can disrupt the balance of city performance. A framework of urban disaster resilience indicators is needed as an instrument to measure the performance of cities in Indonesia in overcoming and adapting to disaster events that can occur anytime and anywhere. From the results of the analysis based on a review of the literature theory and concept of disaster resilience, Regional Capacity Index as an indicator framework is still less able to describe the actual performance conditions of the urban disaster resilience according to the facts of the field. There are physical and social characteristics of the region as attributes of vulnerability to be factors that should also be considered as an indicator of the urban's resilient performance against the threat of disaster. The formulation of indicator frameworks is proposed which is divided into three main dimensions: the physical-ecological dimension of the region, the socio-economic dimension of the community, and the administrative dimension. The recommendations are expected to be an input for the government and especially to *Badan Nasional Penanggulangan Bencana* (BNPB) or the National Disaster Management Agency (NDMA) in developing a framework of disaster resilience city performance indicators in Indonesia.

Keywords Indicator frameworks · Indonesian cities · Regional capacity index · Urban disaster resilience

R. Pranadi (✉) · L. Yola
Department of Urban Studies, School of Strategic and Global Studies, Universitas Indonesia, Jakarta, Indonesia
e-mail: raditya.pranadi81@ui.ac.id

T. Nurlambang
Department of Geography, Faculty of Mathematics and Natural Sciences, Universitas Indonesia, Depok, Indonesia

L. Yola et al. (eds.), *Sustainable Architecture and Building Environment*, Lecture Notes in Civil Engineering 161,
https://doi.org/10.1007/978-981-16-2329-5_8

1 Introduction

City is a manifestation of the progress of civilization in a society because it is formed from human activities, create interactions between spaces and form a spatial system that causes the emergence of the urbanization process. Indonesia's urban population in 2045 will reach 67.1%. Meanwhile, Indonesia is one of the countries with the highest disaster risk due to its geographic position which is between the junction of the three tectonic plates of the world and is included in the Inter-tropical Convergence Zone (ITCZ). Indonesia is the country in the world most exposed to natural disasters with around 60% of the total population in 2018 at risk [1, 2].

Measuring appropriate indicators, such as through a multi-risk understanding approach, helps city governments assess current and future needs and capacities so that cities can develop comprehensive strategies [3]. The National Disaster Management Agency or *Badan Nasional Penanggulangan Bencana* (BNPB) has a set of assessment indicators namely Regional Capacity Index (RCI) as a measuring tool for regional readiness to face the threat of disaster which is then proxied as a measure of the level of resilience of a province/district/city. However, the RCI still focuses on the city's ability to overcome threats which is measured by the presence (or absence) of structural and/or non-structural mitigation policies or activities. Meanwhile, resilience is an attribute in a social system or place with different characteristics of each city and can face a variety of disturbances. This study seeks to find out the framework of measuring the urban disaster resilience appropriately in the context of regional conditions in Indonesia.

2 Research Methodology

This research was conducted through a descriptive-qualitative approach. To understand disaster risk management, including disaster resilience well use the system approach. Due to the Corona Virus Disease 2019 (COVID-19) pandemic, the research method is then carried out an exploratory study and in-depth literature review through the study of scientific articles and other literature sources. This research focuses on Indonesia by taking several cities as sample case study locations to help explain the analysis. The selection of several sample cities considers the historical background of disaster events that have been experienced in the city as well as the availability of data owned. Out of a total of 98 cities in Indonesia, only 13 cities have different RCI values a year.

3 Results and Discussion

The calculated RCI value data for each region is obtained through the BNPB's InaRISK application which can be accessed through the inarisk.bnpb.go.id online page. However, the available RCI data for each city is incomplete for all regions because not all local governments conduct the assessment at the same time. Therefore, out of a total of 98 cities in Indonesia, only 13 cities have different RCI values a year. Therefore, the analysis of urban disaster resilience performance was carried out only in 13 cities as follows in Table 1.

To find out the performance of the city's disaster resilience, an analysis of RCI value increase obtained to the data of disaster events that occurred. Cities are grouped by disaster type so that they can be comparatively analyzed between cities with the same characteristics of disaster events to obtain the right interpretation. The 13 identified cities can be analyzed for their resilience based on disaster type groups of floods and earthquakes.

In the category of flood disasters, there are the cities of Jakarta, Semarang, and Bima that have quite intensive flood disaster events often occur with a record of having a large impact of loss and damage. Two of three cities namely Jakarta and Bima has experienced a decrease in the number of victims affected by the disaster since 2016. This is in line with the increase in their RCI value experienced. However, there are anomalous conditions for another one, Semarang. The number of flood disaster events in 2015 and 2019 decreased compared to the previous year, but there was an increase in the number of victims affected (Fig. 1). If reviewed geographically, the location of the 2019 flood disaster is close to the sea and exposed to flood rob with conditions have a land surface that has decreased and has a high risk of disaster. While the location of the 2015 disaster is mostly in the central area of the city and has a low risk of disaster.

Table 1 Comparison of RCI values in 13 cities in Indonesia

No	City	2016 RCI value	2019 RCI value
1	Jakarta	0.44 (Moderate)	0.75 (Moderate)
2	Semarang	0.49 (Moderate)	0.61 (Moderate)
3	Yogyakarta	0.51 (Moderate)	0.55 (Moderate)
4	Cilegon	0.32 (Low)	0.55 (Moderate)
5	Denpasar	0.50 (Moderate)	0.52 (Moderate)
6	Mataram	0.49 (Moderate)	0.68 (Moderate)
7	Bima	0.49 (Moderate)	0.72 (Moderate)
8	Kupang	0.46 (Moderate)	0.54 (Moderate)
9	Palangkaraya	0.41 (Moderate)	0.54 (Moderate)
10	Samarinda	0.40 (Moderate)	0.50 (Moderate)
11	Gorontalo	0.27 (Low)	0.64 (Moderate)
12	Ambon	0.47 (Moderate)	0.53 (Moderate)
13	Ternate	0.42 (Moderate)	0.59 (Moderate)

Source InaRISK BNPB, accessed in December 2020

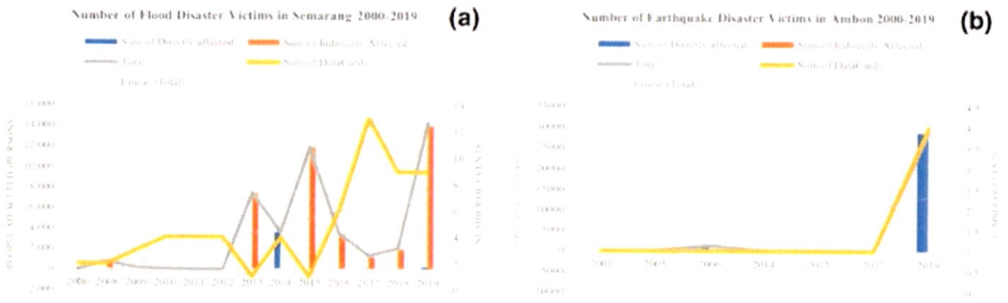

Fig. 1 Number of **a** flood disaster victims in Semarang and **b** earthquake disaster victims in Ambon in 2000–2019

In the category of earthquake disasters, BNPB recorded from 2000 to 2019, many earthquake events occurred in the Bima and Ambon. The increase in RCI value in Ambon shows disharmony with the data on the number of victims affected because the number of victims affected has increased and along with the number of earthquake disaster events that occurred increased. From the interpretation, it was obtained that there are also inconsistencies between the value of RCI and the condition of the city's resilience according to the data in the field. If reviewed geographically, the earthquake disaster in Ambon in 2019 has a large number of affected victims because the epicenter location is on land and close to urban areas. Meanwhile, disasters in previous years were generally more common at sea and away from urban areas (Fig. 2).

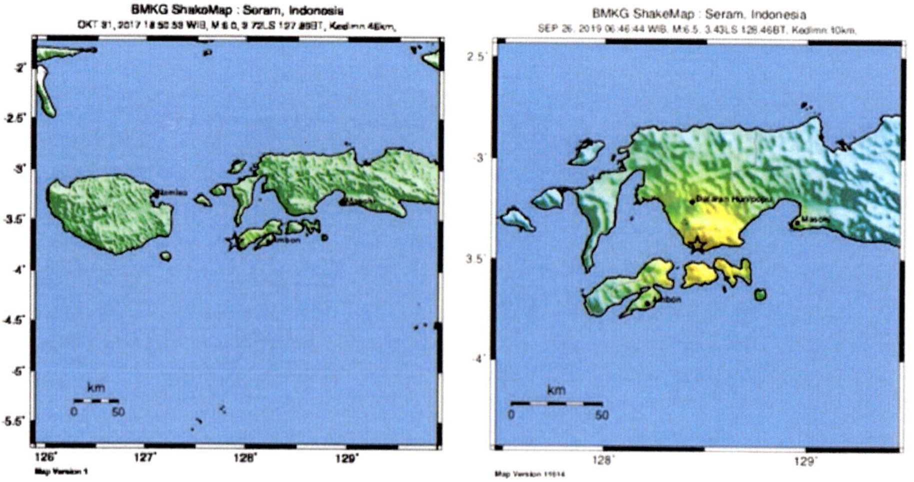

Fig. 2 Comparison of earthquake shakemap in Ambon City in 2017 (left) and 2019 (right) indicates the difference in epicenter location

Table 2 Recommendation of components and variables for developing indicator framework of disaster resilient cities in Indonesia

Components	Variables	Justification
Physical-ecological dimensions		
Land use	% Area of critical land, % area of urbanization (developed land)	[4, 5]
Hazard magnitude	Rainfall, earthquake magnitude, tsunami wave inundation, sea wave height	[4]
Hazard intensity	The average frequency of hazard occurrence	[6]
Topography	% Slope	[5]
Disaster mitigation infrastructure	Availability of disaster mitigation buildings, access distance	[7]
Socio-economic dimension		
Income level	Gini index/Coefficient	[7, 8]
Population	Number of potential exposures people, gender ratio	[6, 9]
Asset ownership	Land/housing ownership (%)	[4, 10]
Education	Level of education, number of schools safe from disaster (disaster curricula)	[4]
Administrative dimension		
Preparedness	Level of public participation, response team trained, disaster literacy, level of understanding of risks in community	[4, 8, 11, 12]
Financial capacity	% Budget allocation for disaster risk reduction activities, % disaster funds	[4]
Planning	Disaster risk reduction-based policies	[12, 13]
Governance and regulation	Effectiveness of disaster early warning systems, regulations	[4]

Source Compiled from any source, 2020

The results of an in-depth analysis of the cases of Semarang and Ambon City showed that there are influential factors in city's resilience, such as the characteristics of disaster events (frequency, duration, intensity) and the nature of vulnerability (distance from the source of danger, conditions of settlements, population density) (Table 2).

Various concepts of disaster resilience indicator frameworks have been compiled and developed by various countries and experts. Based on analysis of various theories, concepts, and data that have been done before, 3 dimensions of performance of the city's resilience to disasters will be formed by a collection of composite indicators, namely the physical-ecological dimension, the socio-economic dimension, and the administrative dimension.

4 Conclusion

It is obtained that the factor that shapes the urban disaster resilience is not only the capacity of the community, but also the physical and social characteristics. RCI has still not been able to provide a comprehensive picture of how the city's resilience to disaster was formed so that repairs are needed. Therefore, three-dimensional performance of the city's resilience to disasters is proposed, which will be formed by a series of composite indicators, namely the physical-ecological dimension, socio-economic dimension, and administrative dimension.

Acknowledgements The preparation of this research certainly cannot be completed properly without the support and constructive input from people around. I thank to Dr. Triarko Nurlambang, M.A. and Dr. Lin Yola, M.Sc. as the supervisor who has provided a lot of guidance and direction in compiling this research. Hopefully this research can be useful for the development of urban management in Indonesia.

References

1. Davis I (2014) Disaster risk management in Asia and the Pacific. In: Disaster risk management in Asia and the Pacific. https://doi.org/10.4324/9781315761336
2. Gunawan I, Sagala S, Amin S, Zawani H, Mangunsong R (2015) City risk diagnostic for urban resilience in Indonesia. World Bank, Jakarta
3. Komendantova N, Scolobig A, Garcia-Aristizabal A, Monfort D, Fleming K (2016) Multi-risk approach and urban resilience. Int J Disaster Resilience Built Environ. https://doi.org/10.1108/IJDRBE-03-2015-0013
4. Shaw R, Takeuchi Y, Joerin J, Fernandez G, Tjandradewi B, Irawati C, Matsuoka Y (2010) Climate and disaster resilience initiative capacity-building program—UNISDR. Kyoto University, Kyoto
5. Saleh TR, Setiadi H (2020) Resilience of flood disasters in the Wanggu Watershed, Kendari City. IOP conference series: earth and environmental science. https://doi.org/10.1088/1755-1315/436/1/012016
6. Manyena B, Machingura F, O'Keefe P (2019) Disaster resilience integrated framework for transformation (DRIFT): a new approach to theorising and operationalising resilience. World Dev. https://doi.org/10.1016/j.worlddev.2019.06.011
7. Cutter SL, Burton CG, Emrich CT (2010) Disaster resilience indicators for benchmarking baseline conditions. J Homel Secur Emerg Manage. https://doi.org/10.2202/1547-7355.1732
8. Norris FH, Stevens SP, Pfefferbaum B, Wyche KF, Pfefferbaum RL (2008) Community resilience as a metaphor, theory, set of capacities, and strategy for disaster readiness. Am J Community Psychol. https://doi.org/10.1007/s10464-007-9156-6
9. Cutter SL, Barnes L, Berry M, Burton C, Evans E, Tate E, Webb J (2008) A place-based model for understanding community resilience to natural disasters. Glob Environ Change. https://doi.org/10.1016/j.gloenvcha.2008.07.013
10. Moghadas M, Asadzadeh A, Vafeidis A, Fekete A, Kötter T (2019) A multi-criteria approach for assessing urban flood resilience in Tehran, Iran. Int J Disaster Risk Reduction. https://doi.org/10.1016/j.ijdrr.2019.101069
11. Baba K, Nagata Y, Kawakubo S, Tanaka M (2020) A framework and indicators of resilience. Resilient Policies Asian Cities. https://doi.org/10.1007/978-981-13-8600-8_1

12. Oktari RS, Comfort LK, Dwitama P (2020) Measuring coastal cities' resilience toward coastal hazards: instrument development and validation. Prog Disaster Sci. https://doi.org/10.1016/j.pdisas.2019.100057
13. Parsons M, Glavac S, Hastings P, Marshall G, McGregor J, McNeill J, Morley P, Reeve I, Stayner R (2016) Top-down assessment of disaster resilience: a conceptual framework using coping and adaptive capacities. Int J Disaster Risk Reduction. https://doi.org/10.1016/j.ijdrr.2016.07.005

How Does Urban Development Contributes to Urban Heat Island: A Decade Increase of Urban Heat Intensity in Jakarta Metropolitan Area

Sofi Ulfiasari and Lin Yola

Abstract Rapid Population growth and economic movement have had a significant impact on the social and environmental change of Jakarta. The dense development of Jakarta goes horizontally and vertically in the city center. The trend also expands and encourages the interaction between the city center and the buffer cities within the Jakarta Metropolitan Area. As a result of the intense socio-economic activities in the city center, the phenomenon of Urban Heat Island (UHI) emerges as the city issue. The rapid increase of UHI intensity worsen the and climate change and is a challenge for Jakarta in realizing a sustainable city agenda. This study indicates a lacking of the focused investigation on the correlation between urban development and UHI in Jakarta Metropolitan area. Therefore, this study aims investigates a decade by using the analysis of Landsat 5 and 8 satellite imagery asses the building density index value and albedo. The variables include the land surface temperature, vegetation index, land cover and satellite light imagery. This study pinpoints the significant impact of urban development on the UHI intensity. The results of this study provide scientific contributions to academia, industry and city governments in the formation of a policy framework for the UHI and climate change mitigation agenda in Jakarta.

Keywords Albedo · Building density index · Jakarta metropolitan urban development · UHI

1 Introduction

Jakarta Metropolitan Area, locally known as 'Jabodetabek' dated from 1997 with strengthening and concentration of the economic concept is the capital city that, handle 50% of circulation of state finances [1] and now became a largest

---

S. Ulfiasari (✉) · L. Yola
Department of Urban Studies, School of Strategic and Global Studies, Universitas Indonesia, Jakarta, Indonesia
e-mail: sofi.ulfiasari91@ui.ac.id

L. Yola et al. (eds.), *Sustainable Architecture and Building Environment*, Lecture Notes in Civil Engineering 161,
https://doi.org/10.1007/978-981-16-2329-5_8

metropolitan area in Indonesia with 6 cities in 3 provinces. In 2019, the Jakarta population a central of Jabodetabek has reached 10.5 million [2]. This trend raises environmental issues, especially the increase in urban air temperature, which is one of the main contributors to climate change. In this context, one of the focus issues that need to be targeted for a solution in Jakarta is the Urban Heat Island (UHI) phenomenon. The temperature increase in the city area significantly affects the level of thermal comfort and the energy consumption. A 0.03 °C yearly temperature increase accelerates atmospheric modification conditions [3].

UHI is defined as the temperature increases in urban areas compared to the surrounding environment that starts after sunset [4–6]. UHI can achieve a maximum temperature rise of 5.6 °C in urban zones [7, 8]. While the temperature difference between urban and rural can reach 12 °C in a metropolitan city [9]. Jakarta's surface temperature can exceed 34 °C which reaches the highest level 3–4 h after sunset in downtown Jakarta [10]. Jakarta experienced the highest temperature increase (0.82 °C). In 2019, the year that experienced the second highest temperature rise after the 2016 phenomenon, compared to the temperature vulnerability of 1981–2010. This illustrates that UHI is a significant issue that requires urgent solutions in Jakarta Metropolitan Area, because the UHI issue is one of the main challenges in realizing the climate change mitigation agenda. If this phenomenon continues, it will give an impression of an increase in global temperatures [11].

The factors that cause UHI [12] are increased anthropogenic heat, reduced evaporation, increased heat storage, increased radiation and reduced convection. In addition, population and urban density measures will also affect the intensity of the urban heat island [7]. Asian cities are growing with increasing population and city size rapidly, due this situation there is any variation and trend changing in urban heat island [13]. Populations may also track the trailing climate through migration or dispersal, or they may go extinct [14].

The increase in the intensity of the UHI is a variable that cannot be separated from the discussion of climate change mitigation. Reducing increased emissions is one of the national targets on the agenda in the National Action Plan for Reducing Greenhouse Gas Emissions (RAN-GRK), as a follow-up to Indonesia's commitment to dealing with climate change issues. However, there is lacking of the intensive studies on the past decade on UHI in Jabodetabek. Therefore, this study aims to provide a study on the mapping of UHI as a measure to mitigate the intensity of UHI in the Jakarta Metropolitan Area, the largest metropolitan area facing the highest UHI issue in Indonesia today. The use of remote sensing images is able to obtain the intensity and distribution of UHI temporally [15]. So that UHI detection can be done in a relatively short time compared to having to measure in the field [16].

2 Literature Review

2.1 Urban Climate

Urban climate is the result of the interaction between the urban surface and the lower atmospheric layer which takes place radiatively, thermodynamically, and aerodynamically so that it can change surface temperature, air temperature, humidity, rainfall, air quality, surface energy flux, mixing layer height, boundary layer wind, and turbulence in urban areas [17]. Urban Heat Island is a characteristic of high temperature that is concentrated in an urban area [16]. Urban Heat Island arises because the isotherm pattern forms an island-like pattern. The temperature difference between urban and rural can reach 12 °C in a metropolitan city [9]. Urban Heat Island occurs as a result of changes in temperature, albedo, evapotranspiration, and energy flux [12]. This phenomenon is a clear example of humans being careless in modifying the climate [18]. According to Oke [7] and Sobrino et al. [19], the spatial and temporal characteristics of UHI vary widely depending on local changes in urban form and function. The size of the UHI is also strongly influenced by local meteorological conditions and geography (topography, presence of water bodies such as lakes or rivers, soil types, etc.) of the local area. In addition, population and urban density measures will also affect the intensity of the UHI [7].

Microclimate scale urban heat island have been frequently associated with global warming [20]. The urban heat island effect resembles a system that makes urban residents need air conditioning, while air conditioning will increase electricity consumption and can contribute to urban heat. Some of the negative effects of urban heat island among them the deaths of hundreds of people in the summer caused by heat wave [21]. The energy consumption is high in the area with high UHI variations and the other way around [22]. The rising greenhouse gas concentration cause an elevation in temperature at both urban and less urbanized area direct urban heat island to increase [23]. Currently, human activities are no longer limited to only in the morning, technology has helped humans to move into the night, this has led people, especially in urban areas, to use it to continue producing even up to 24 h [24].

2.2 Urban Development

With the background of rapid urbanization, the population living in urban areas is forecasted to be 5 billion by 2030 [25]. Urban development cause urbanization and migration of population from rural to urban areas, this phenomenon isolated population centres are changing into metropolitan cities and still growth because urban development was viewed as an end result of human actions, and the value system of urban society as the primary source of the impulse for actions [26] Urban sprawl is the phenomenon of disproportional expansion of urbanized areas into undeveloped land [27].

Human activities on earth inevitably affect land cover which refers to the physical and biological cover above the soil surface, including water, vegetation, vacant land and/or artificial structures [28]. Numerous modifications of land surface will occur as an accumulating number of people migrate into metropolitan areas [29]. So that one indicator of the UHI is the amount of waterproof material can store heat [30].

The configuration of spaces and buildings in the dense urban area plays a significant role in the urban microclimate and outdoor activities [31]. The solar radiation trapping effects in urban spaces depend on urban configuration, vertical obstruction and roughness play roles of creating shadow effect [32]. The presence of natural vegetation influenced a lot in the distribution of Land Surface Temperature [33].

3 Methods

The computer simulation is a reliable tool to climatically responsive urban configuration study, technology helped climate study more effective and efficient [34]. Surface Urban Heat Island first observation from satellite-based sensors in 1972 and landsat is the one which has thermal sensor which can provide temperature in real time [16]. The difference daytime and nighttime urban heat island is the anthropogenic heat had bigger effect at night, based from computer simulation [35]. So that, in this study using nightlight imagery satellite to provide how heat produce at night that can increase heat storage. To detect surface urban heat island and its correlation with urban development, built-up index, green index and city light are needed to correlate with surface temperature.

This study was conducted in the area of Jakarta metropolitan area (Jabodetabek) situated at 3 province which are DKI Jakarta, West Java, and Banten. A fullscene Landsat 5 and 8 (path 122 row 64 and 65) of 2009/07/29, 2014/9/13 and 2019/8/1 were acquired with a little cloud cover. The very high resolution (VHR) imagery was used to calculate the percentage of a Normalized Difference Built-up Index and land surface temperature. Nightlight satellite data image from NOAA of 2014/7/1 and 2019/4/30 was used to extract city light.

The formula for calculating LST For Landsat 5,

$$\text{Radiance} = \frac{\text{LMAX} - \text{LMIN}}{\text{QCalmax} - \text{QCalmin}} \times (\text{QCal} - \text{QCalmin}) + \text{LMIN} \quad (1)$$

where the gain and offset can be obtained from the header file, Qcalmin = 1, Qcalmax = 255, Qcal = DN, and Qmax and Qmin.

For Landsat 8 Using the radiance rescaling factor, Thermal Infra-Red Digital Numbers can be converted to TOA spectral radiance.

$$L\lambda = ML * Qcal + AL \tag{2}$$

where

$L\lambda$ = TOA spectral radiance (Watts/(m^2.sr. μm))
ML = Radiance multiplicative Band (No 10)
AL = Radiance Add Band (No 10)
Qcal = Quantized and calibrated standard product pixel values (DN).

Spectral radiance data can be converted to top of atmosphere brightness temperature using the thermal constant Values in Metadata file,

$$T = \frac{K2}{\text{In } (K1/L\lambda) + 1} \tag{3}$$

where T is the effective at-satellite brightness temperature in Kelvin, K1 = 666.09 (m.W.cm^{-2} . r^{-1}) and K2 = 1282.71 K are calibration constants, and TTM6 is the spectral radiance in m.W.cm^{-2} . sr^{-1}.

The formula for calculating Normalized Difference Built-up Index as follows [36]. Simply, in the NDBI image the built-up area and barren pixels have a positive values, whereas the negative or zero values mean other land covers, allowing the built-up regions to be set automatically.

$$NDBI = \frac{\text{SWIR} - \text{Nir}}{\text{SWIR} + \text{Nir}} \tag{4}$$

Normalized Difference Vegetation Index shown biomass and vegetation health index [37] which has correlation with land surface temperature [38]. The formula of NDVI has result from −1 to +1, means no vegetation to highest possible density of green leaves [39]

$$NDVI = \frac{\text{Nir} - Red}{\text{Nir} + Red} \tag{5}$$

To understanding the correlation between land surface temperature, build-up area and city light, the approach is used to estimate correlation on the linear regression model as follows:

$$Y = a + bx \tag{6}$$

4 Result and Discussion

4.1 Urban Development

Being the capital city of Indonesia, the sprawling metropolitan city of Jakarta is the centre of government as well as business activities in Indonesia. Jabodetabek is a metropolitan district dated 1997 became the fastest growing city in Indonesia and the world's second-largest conurbation, the development in every city and its integration through infrastructure creates more and more built-up areas. Urbanization occurred in every region, until it spread closer to Jakarta. Data processing Landsat satellite imagery using SWIR and NIr sensors produces built-up index to show how the cities grown.

At last, 10 years development in Jabodetabek, built-up area growth 250% from west to east, high density built-up area growth up to 500% from 2009 to 2019 and found in core cities (Fig. 1). Built-up area growth following the infrastructure and access, the low density shown increasing new housing in the suburbs. The high density built-up means more areas covered by high buildings which can indicate as a characteristic of a metropolitan city. The high density building found in North Jakarta at 2009 and grows in Tangerang (westside) until Bekasi (eastside) at 2014–2019. High density built areas grew rapidly in Tangerang and Bekasi, in 2009 there was a low density, then changed to high density in 2014 so that the surrounding area turned into a built area with low density (Table 1).

National Oceanic and Atmospheric Administration produce nightlight satellite imagery that can shown city light as a product of human activities at night (city light) shown in Fig. 2. City light comes from the use of lights in built-up areas, lights that come from vehicles and objects that radiates heat.

In 2014, there were high brightness reaching 640 nW/cm^2/sr, but in 2019 the highest brightness level was only 385 nW/cm^2/sr. Figure 2 shown that Jakarta as a core and the city light is fading away from the centre, straight line pattern (road)

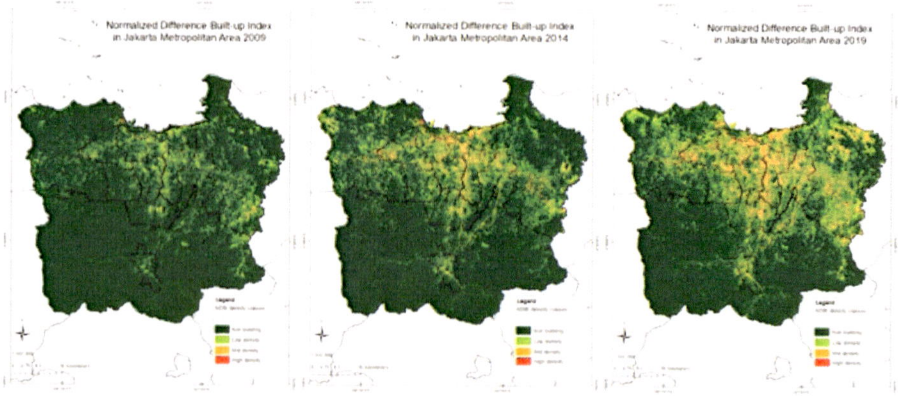

Fig. 1 Jakarta metropolitan area normalized difference built-up index 2009, 2014 and 2019

Table 1 Normalized difference built-up index area

No	Value	Dense index	km^2 (2009)	km^2 (2014)	km^2 (2019)
1	<0	Non building	5855.17	5246.40	4406.74
2	0–0.1	Low density	795.57	1080.14	1518.74
3	0.1–0.2	Mid density	153.773	445.10	840.14
4	0.2–0.3	High density	9.45	42.23	47.52

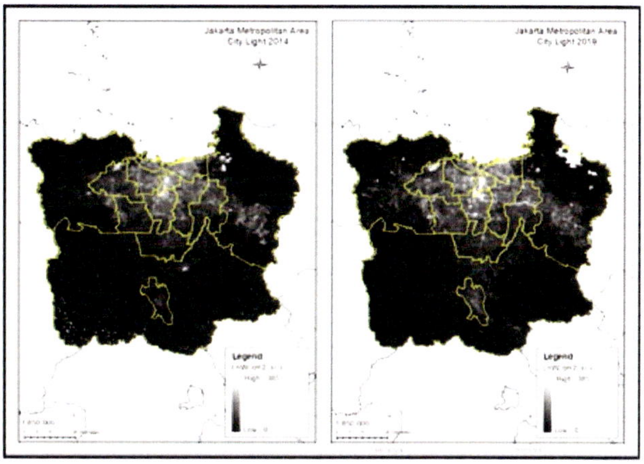

Fig. 2 Jakarta metropolitan area city light 2014 and 2019

with high brightness from the centre to the east. In 2019, city lights appear stronger and flatter, the light connected each region and the most brightness is in each core. In 2014, the highest city light appears when traffic jam at the core of the city but in 2019, city light spread up rapidly from the core (Jakarta) to the suburbs (west-east).

Spatially, NDVI that is shown in Fig. 3 in Jakarta Metropolitan Area non vegetation found in coastal areas, in 2009 the data was taken at harvest time so that

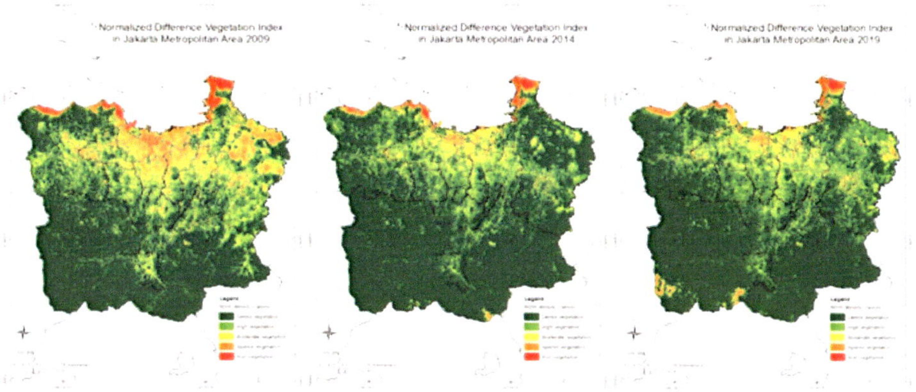

Fig. 3 Jakarta metropolitan normalized difference vegetation index 2009, 2014 and 2019

Table 2 Normalized difference vegetation index area

No	Value	Dense index	km^2 (2009)	km^2 (2014)	km^2 (2019)
1	<−0.03	Non vegetation	90.64	134.21	106.39
2	−0.03–0.15	Sparse vegetation	611.45	222.95	267.62
3	0.15–0.25	Moderate vegetation	867.37	647.97	818.43
4	0.25–0.35	High vegetation	866.71	830.67	1060.61
5	0.35–1	Dense vegetation	4341.68	5023.27	4606.04

the level of greenness was relatively low even though the three data were recorded during the dry season (Table 2).

4.2 *Urban Heat Island*

Jakarta Metropolitan Area LST distribution ranges from 0 to 40 °C with a centralized pattern in Jakarta and evenly spreads to its suburbs. Urban heat island occurs as a result of urban growth that is getting faster and wider in suburban areas. In 2019, the temperature distribution approaching 40 °C is in the Jakarta area as a whole and there is a line pattern that stretches from south to north that divides the centre of Bogor City of Jakarta according to the main road (Fig. 4).

The temperature value in suburban areas ranges from 18–34 °C, sub-urban areas that have a relatively flat altitude and the same as Jakarta have a temperature difference of up to 5 °C, while for areas that have a different altitude, the temperature difference reaches 10 °C. When viewed from the distribution pattern, the highest temperatures occur in areas with the highest human activity such as in areas

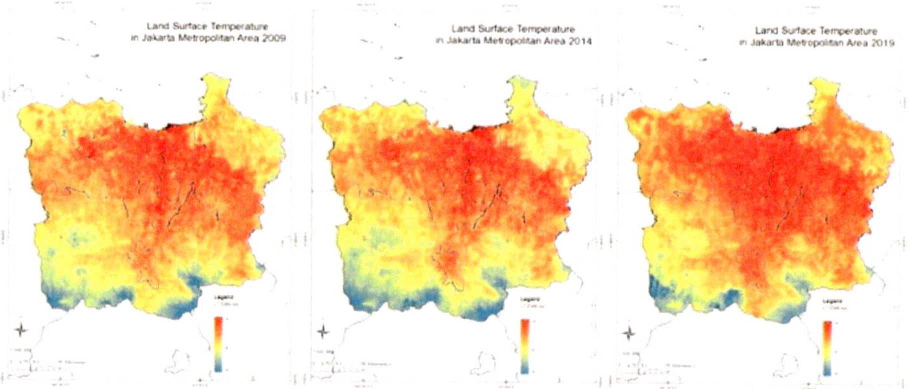

Fig. 4 Jakarta metropolitan area land surface temperature in 2009, 2014 and 2019

Table 3 Land surface temperature's correlation

Correlation	2009	2014	2019
LST and NDVI	LST = −8.0566 NDVI + 32.687 (r^2 = 0.3108)	LST = −8.0495 NDVI + 32.687 (r^2 = 0.3248)	LST = −8.6952 NDVI + 34.559 (r^2 = 0.2574)
LST and NDBI	LST = 13.514 NDBI + 31.921 (r^2 = 0.5058)	LST = 15.13 NDBI + 30.68 (r^2 = 0.5137)	LST = 17.173 NDBI + 31.672 (r^2 = 0.4601)
LST and City Light		LST = 0.2864 City Light + 27.061 (r^2 = 0.4417)	LST = 0.1865 City Light + 28.687 (r^2 = 0.3847)

around transportation routes, downtown areas and industrial areas. The minimum and maximum land surface temperature in Jabodetabek increase 3–5 °C at last 10 years (Table 3).

Correlation analysis has been done to find out the relationship between LST and NDVI which have shown a negative correlation, land surface temperature will lower if the dense vegetation is high. The city lights only have a small impact as temperatures at night, this variable can show how the cities works for 24 h, which can contribute to warming when the UHI occurs at night. Electricity and vehicle consumption at night shows a correlation with temperature. However, conditions that are too small in scale cannot provide detailed data so it is necessary to calculate electricity consumption and exhaust gas results in vehicles at night. NDBI and City light are the factors that influence the land surface temperature seen in the positive correlation results. NDBI correlated by 50% and City light correlated by 40% in increasing land surface temperature, while NDVI has a negative correlation means when NDVI increases, the temperature will decrease.

5 Conclusions

The urban temperature in Jabodetabek has reached over 30 °C is a real threat to not only city thermal comfort but also the energy consumption. This study points out that there is a rise of UHI intensity from 2009 to 2019 in the past decade. Economic activity is the biggest cause of accelerated temperature rise, changes in land cover make heat storage more widespread. City lights at night only account for 20% of the increase in temperature. Concentrations of City light brightness occur in City centers and streets (straight line) indicates the traffic jam, that may cause the temperature rising. Built-up area increasing 500% in a decade from west to east, the highest density found at North Jakarta, Tangerang and Bekasi. This phenomenon will make the urban climate system rapidly change. Built-up area increasing electrical consumption that increasing city light and heat storage in city, when the heat sorage larger than the ability of vegetation to absorb heat. So that vegetation in city can't withstand increasing land surface temperatures. The finding of this study is a significant reference for UHI and climate change mitigation national agenda to enhance the movement towards the cities sustainability targets.

References

1. Hendaru Tri Hanggoro (2018) Tujuan Konsep Jabotabek Meleset? Historia. https://historia.id/urban/articles/tujuan-konsep-jabotabek-meleset-vQNZ8
2. BPS-Statistics Indonesia (2019) Jakarta Dalam Angka 2019
3. Meteorological, Climatological, and G. A. B (2019) Tren Suhu. https://www.bmkg.go.id/iklim/?p=tren-suhu
4. Erell E (2012) Urban microclimate. In: Urban microclimate. Routledge. https://doi.org/10.4324/9781849775397
5. Oke TR (1988) The urban energy balance. Prog Phys Geogr 12(4):471–508. https://doi.org/10.1177/030913338801200401
6. Van Bohemen H (2012) (Eco)System thinking: ecological principles for buildings, roads and industrial and urban areas. In: van Bueren E, van Bohemen H, Itard L (eds) Sustainabl. Springer, Dordrecht. http://doi-org-443.webvpn.fjmu.edu.cn/10.1007/978-94-007-1294-2_2
7. Oke TR (1973) City size and the urban heat island. Atmospheric environment. https://doi.org/10.1016/0004-6981(73)90140-6
8. Wang Y, Li Y, Di Sabatino S, Martilli A, Chan PW (2018) Effects of anthropogenic heat due to air-conditioning systems on an extreme high temperature event in Hong Kong. Environ Res Lett. https://doi.org/10.1088/1748-9326/aaa848
9. Oke TR (1997) Urban climates and global environmental change. In: Thompson RD, Perry A (eds) Applied climatology: principles and practices. Routledge, New York, NY
10. Rushayati SB, Hermawan R (2013) Characteristics of urban heat island condition in DKI Jakarta. Media Konservasi
11. Fallmann J, Emeis S (2020) How to bring urban and global climate studies together with urban planning and architecture? Dev Built Environ 4:100023. https://doi.org/10.1016/j.dibe.2020.100023
12. Gartland L (2012) Heat islands: understanding and mitigating heat in urban areas. https://doi.org/10.4324/9781849771559
13. Lee K, Kim Y, Sung HC, Ryu J, Jeon SW (2020) Trend analysis of urban heat island intensity according to urban area change in asian mega cities. Sustainability (Switzerland). https://doi.org/10.3390/su12010112
14. Franks SJ, Weber JJ, Aitken SN (2014) Evolutionary and plastic responses to climate change in terrestrial plant populations. Evol Appl. https://doi.org/10.1111/eva.12112
15. Fawzi NI (2017) Measuring urban heat island using remote sensing. Case of Yogyakarta City, Majalah Ilmiah Globe
16. Voogt JA, Oke TR (2003) Thermal remote sensing of urban climates. Remote Sens Environ. https://doi.org/10.1016/S0034-4257(03)00079-8
17. Arya S (2001) Introduction to microclimatology. In: 2nd ed. Academic Press
18. Oke Timothy R, Mills G, Christen A, Voogt JA (2017) Urban climates. Urban Climates. https://doi.org/10.1017/9781139016476
19. Sobrino JA, Oltra-Carrió R, Sòria G, Bianchi R, Paganini M (2012) Impact of spatial resolution and satellite overpass time on evaluation of the surface urban heat island effects. Remote Sens Environ 117:50–56. https://doi.org/10.1016/j.rse.2011.04.042
20. Alcoforado MJ, Andrade H (2008) Global warming and the urban heat island. In: Urban ecology: an international perspective on the interaction between humans and nature. https://doi.org/10.1007/978-0-387-73412-5_14

21. Oudin Åström D, Bertil F, Joacim R (2011) Heat wave impact on morbidity and mortality in the elderly population: a review of recent studies. Maturitas. https://doi.org/10.1016/j.maturitas.2011.03.008
22. Arifwidodo S, Chandrasiri O (2015) Urban heat island and household energy consumption in Bangkok, Thailand. Energy Procedia. https://doi.org/10.1016/j.egypro.2015.11.461
23. Sachindra DA, Ng AWM, Muthukumaran S, Perera BJC (2016) Impact of climate change on urban heat island effect and extreme temperatures: a case-study. Q J Royal Meteorol Soc. https://doi.org/10.1002/qj.2642
24. Henckel D (2018) The 24/7 City-a Myth? https://doi.org/10.13140/RG.2.2.30335.71842
25. Ash C, Jasny BR, Roberts L, Stone R, Sugden AM (2008) Reimagining cities. Science. https://doi.org/10.1126/science.319.5864.739
26. Liu Y (2008) Modelling urban development with geographical information systems and cellular automata. In: Modelling urban development with geographical information systems and cellular automata. https://doi.org/10.1201/9781420059908
27. Zhang T (2000) Land market forces and government's role in sprawl. The case of China. Cities. https://doi.org/10.1016/S0264-2751(00)00007-X
28. EPA (2008) Urban heat island basics. In: Reducing urban heat islands: compendium of strategies| heat island effect. US EPA
29. Wang K, Wang J, Wang P, Sparrow M, Yang J, Chen H, Wang C (2007) Influences of urbanization on surface characteristics as derived from the moderate-resolution imaging spectroradiometer: a case study for the Beijing metropolitan area. J Geophys Res 112: D22S06. https://doi.org/10.1029/2006JD007997
30. Yuan F, Bauer ME (2007) Comparison of impervious surface area and normalized difference vegetation index as indicators of surface urban heat island effects in Landsat imagery. Remote Sens Environ. https://doi.org/10.1016/j.rse.2006.09.003
31. Yola L (2018) Impact of urban configurations on microclimate and thermal comfort in residential area of Kuala Lumpur
32. Yuan X, Wang W, Cui J, Meng F, Kurban A, De Maeyer P (2017) Vegetation changes and land surface feedbacks drive shifts in local temperatures over Central Asia. Scientific Reports 7(1):3287. https://doi.org/10.1038/s41598-017-03432-2
33. Yola L, Siong HC, Djaja K (2020) Climatically responsive urban configuration in residential area: research gaps. AIP Conf Proc 2020(2255):0700141
34. Yola L, Siong HC (2017) Computer simulation as an alternative approach in climatically responsive urban configuration study. Chem Eng Trans 56:505–510
35. Atkinson BW (2003) Numerical modelling of urban heat-island intensity. Bound-Layer Meteorol. https://doi.org/10.1023/A:1025820326672
36. Zha Y, Gao J, Ni S (2003) Use of normalized difference built-up index in automatically mapping urban areas from TM imagery. Int J Remote Sens 24:583–594. https://doi.org/10.1080/01431160304987
37. Aparicio N, Villegas D, Araus JL, Casadesús J, Royo C (2002) Relationship between growth traits and spectral vegetation indices in durum wheat. Crop Sci. https://doi.org/10.2135/cropsci2002.1547
38. Zhang Yanbin AN, Nan Liu Peiyan et al (2017) An analysis of land surface temperature (LST) and its influencing factors in summer in western Sichuan Plateau: a case study of Xichang City. Remote Sen Land Resour 29(2):207–214. https://doi.org/10.6046/gtzyyg.2017.02.30
39. Observatory E (2011) Measuring vegetation (NDVI & EVI). Earth observatory. https://earthobservatory.nasa.gov/features/MeasuringVegetation/measuring_vegetation_2.php#: ~ :text=Measuring Vegetation (NDVI %26 EVI)&text = To determine the density of,up the spectrum of sunlight

Urban Self-organization from Approaching Collective Home Spheres, Through the Case of Kampung Akuarium, Jakarta

Hiroto Ota

Abstract This paper proposes an original method of group dialogue with inhabitants, called Sketch-log Dialogue, that combines Bohm Dialogue and free drawing, with the goal of developing an approach to planning and design that fosters the urban self-organization. This method was put into practice in Kampung which is said to have a high capacity for self-organization. As a result, a variety of Home spheres were drawn in the kampung Akuarium depending on the values and experiences of each individuals, and the inherent order and organization by the community were revealed. This method is useful for improving the living environment in detail by focusing on the individual as the smallest unit of self-organization.

Keywords Self-organization · Home sphere · Dialogue · Kampung · Urban planning

1 Introduction

1.1 Planning Theory to Foster Urban Self-organization

Since the end of the twentieth century, the breakdowns in modern urban planning have been pointed out one after another, and the concept of urban self-organization has been gaining attention. This is a key mechanism through which cities spontaneously and dynamically evolve and adjust to changing circumstances [1–3]. In parallel, there has been a rapid increase in articles and discourses [4]. Self-organization is a term used by researchers in a variety of ways, but the main stream in urban studies is to encapsulate theories and concepts of informal planning, insurgent planning, active (urban) citizenship, and bottom-up participation in planning [5]. As Marshall [6] observes: "city planning is effectively premised on the idea of the unplanned city as disordered and dysfunctional". In this sense, a slum is

H. Ota (✉)
Department of Architecture, Faculty of Engineering, Musashino University, Tokyo, Japan
e-mail: h-ota@musashino-u.ac.jp

L. Yola et al. (eds.), *Sustainable Architecture and Building Environment*, Lecture Notes in Civil Engineering 161,
https://doi.org/10.1007/978-981-16-2329-5_10

seen as a settlement to be removed. But is it right to judge that the settlement should be removed in a disorderly manner solely because of the lack of physical conditions? It would lead to the loss of the urban cultural landscape and their own residential culture. In Indonesia, there are communities and close settlements called kampung; being mostly a self-initiated and self-constructed urban settlement, with sometimes low housing qualities and always no security of tenure [7]. According to the current urban planning, kampungs are often judged as "slums". The word "slum" is not only associated with the lack of physical conditions, but also with a negative image of the people living there and their lives [8]. In fact, kampungs have capacity to their own condition 'organized' and to be regarded as environmentally continuous and culturally authentic living, even if the density of residents is very high [9]. According to S. Moroni et al. self-organization has the following three characteristics; Self-building, Self-governance, Self-coordination [10]. Kampung can be thought of as a living environment that makes full use of these three abilities and continues to grow in a flexible way. This paper will focus on the smallest scale of self-organization. That is, the design of the living environment by individuals. The three characteristics of self-organization are also based on individual action at the micro level. These actions are based on people's value judgments, independent of public standards, and planners and designers must learn their value systems. Specifically, this paper introduces a drawing-based interaction method developed by H. Ota [11] as an approach to valuing individual residents.

1.2 Case Study in Kampung Akuarium, Jakarta

In 2016, Kampung, on the coast of North Jakarta, was evicted by the Jakarta Provincial Administration under the regime of the former provincial governor for projects aimed at infrastructure development and greening; Kampung Akuarium was one of them, and about 500 families lost their homes (Fig. 1).

However, in April 2017, the current governor of the province took office and changed the policy to return to the settlement, and a temporary shelter was built in

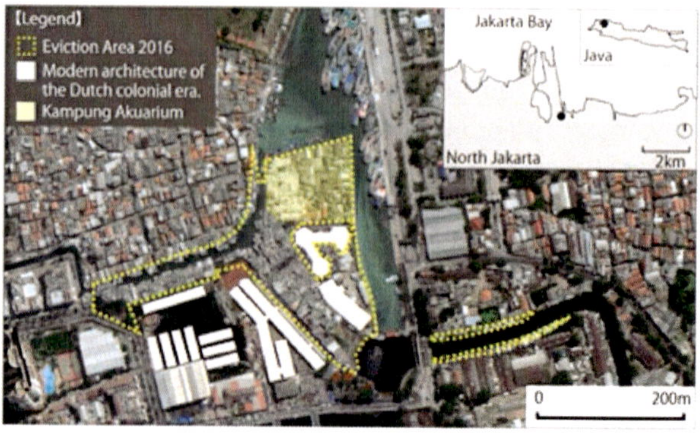

Fig. 1 Site of Kampung Akuarium

2018. About few months after eviction, Rujak Center for Urban Studies, together with several organizations and with the displaced community, has supported and empowered the people in early process of recovery, to recover the citizenship ID, to communicate with the state's related sectors, to coordinate the temporary housing and finally to plan the new housings [9]. In the new plan should be not only the physical environment but also the entire human environment with all of its tangible and intangible qualities. In the past two years, they have organized the shelter in their own way, transforming it from an inorganic shelter into lively living environment. In this paper, we made dialogue with the inhabitants in the temporary shelter for approaching to their own Home sphere and Evaluating by an original way of sketches [11].

2 Method: Sketch-Log Dialogue

① Based on "Bohm Dialogue"

H. Ota has developed the original method of dialogue, called "Sketch-log Dialogue". It is based on the principles of Bohm Dialogue [10], is below 1–4. (1) The group agrees that no group-level decisions will be made in the conversation. (2) Each individual agrees to suspend judgement in the conversation. (3) As these individuals "suspend judgement" they also simultaneously are as honest and transparent as possible. (4) Individuals in the conversation try to build on other individuals' ideas in the conversation. In other words, dialogue is not about saying what each other has thought of beforehand, but about sharing new meanings at that moment and place. Dialogue practitioners do not use operating concepts, but try to elicit stories based on informants' values and value judgments.

② Group dialogue with Sketch-log

The dialogue—to one inhabitant at a time—proceeds from talking about individual life history in the city to talking about what he is evaluating in his own surroundings. At the same time, the participants make 'hand-drawing sketches' of the contents to visualize the notions, share, and confirm (Fig. 2).

You have to draw a sketch and then show it to the informant, if wrong rewrite it, again and again until they completely understand each other by unstructured way. The sketch in this method is not a correct record of the phenomenon, but a scaffold for communication, and the drawn Sketch-log is evidence that the meaning content could be shared among the interlocutors. Drawing and showing sketches to each other at the same time makes communication more interactive. In this method (Fig. 2), approachers usually draw several dozen sketches in one dialogue. Also, the several scales and times of maps help you to communicate easily.

Fig. 2 Sketch-log dialogue

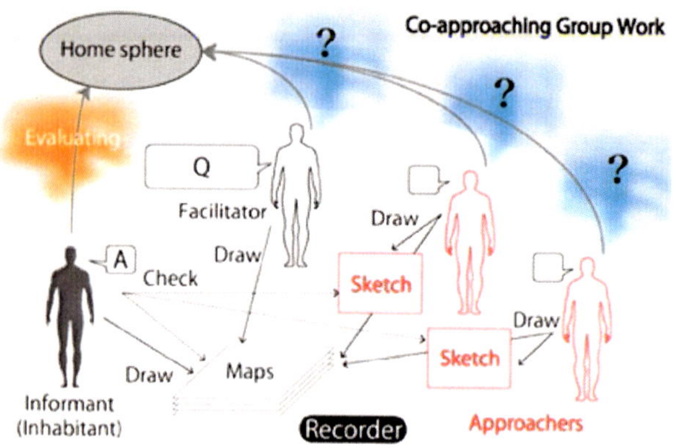

③ Approaching Home sphere in Kampung Akuarium

The Home sphere is not just a range of living activities, but a living environment that is valued by its inhabitants. Therefore, it is more important to understand the value judgments and the mechanisms behind them than the areas and forms. In this practice, we tried to approaching Kampung people's Home sphere, Area dimana yang dirasakan seperti rumah sendini (in Indonesia), and evaluating to their own surroundings in the temporary shelter and also to the living environment before eviction, and their memories. After dialogue, we hold the meeting with the community and cross-checked about our output.

3 Result

3.1 *Inhabitant's Home Sphere and Evaluating*

We made dialogue with 12 inhabitants, varied in gender and livelihood, from Nov 21th to 25th, 2018. The location was arranged by the inhabitants, such as in their house, under the eaves, or Mushola (Table 1).

Table 1 List of informants

Informant	Date	Age	Gender	Block	Note
A	Nov 21/2018	30's	f	C	Housewife
B, C	Nov 22/2018	40's, 40's	f, m	C, C	Shop clerk, shop owner
D	Nov 22/2018	40's	f	A	Housewife
E	Nov 23/2018	30's	m	C	Fisherman
F	Nov 23/2018	30's	f	C	Housewife
G	Nov 23/2018	30's	m	C	Trailor, RT leader
H, I	Nov 24/2018	60's, 30's	f, f	A, A	Housewife
J, K	Nov 25/2018	40's, 40's	f, f	B, B	Aquarium shop, housewife
L	Nov 25/2018	30's	f	C	Housewife

The results obtained are shown in the Fig. 3. The example sketch shown is one of several dozen. As you can see, Home spheres have very diverse shapes based on the values of each inhabitant. It is closely related to personal experiences and memories in this kampung. For example, Informant-A came from Luar Batang, next to Kampung Akuarium, and she's house located near by the steel bridge before eviction. In front of her house, people were gathering before, and she evaluated such a meeting space. She also liked to see pasar ikan before. So, we drew her Home sphere like Fig. 3, it includes not only in Kampung Akuarium but also in Luar Batang. Others, Informant-D, whose home sphere is the road to the neighboring kampung, says the kitchen in front of his house is the most important, and he contributes to the community through the kitchen. Informant-E, a fisherman, saw the Home sphere on the sea and the open space playing soccer with children. L, who says the children of Kampong are the most important, has opened up her house as a second kindergarten. They are actively taking action in the Home sphere and contributing to the living environment by invoking it. In other words, it can be said that the kampung is self-organizing through the Collective Home sphere and evaluating by the inhabitants.

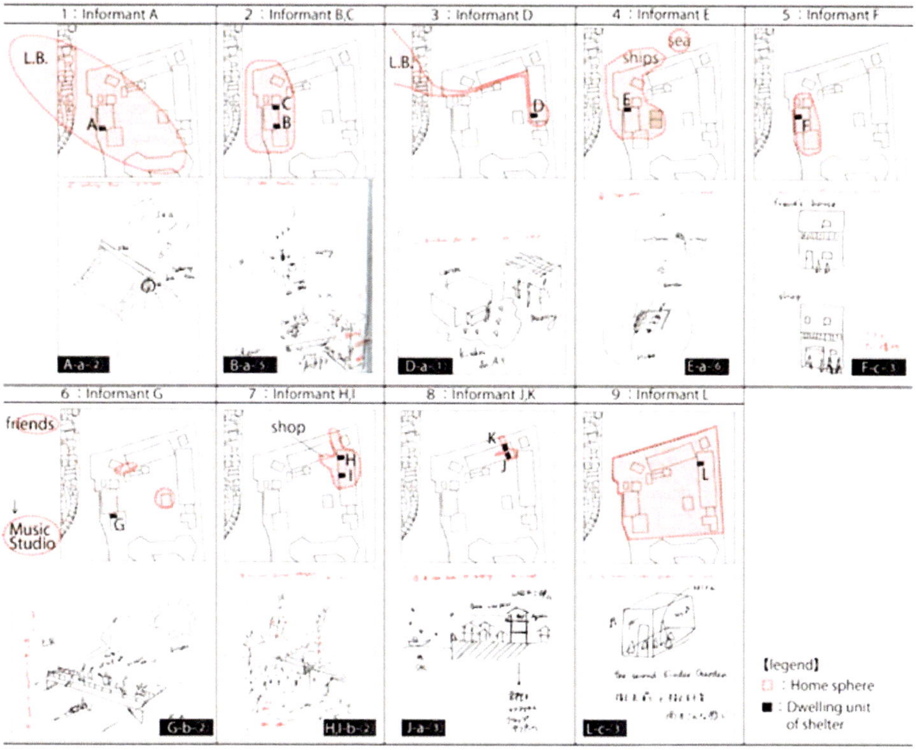

Fig. 3 Collective home sphere and evaluating

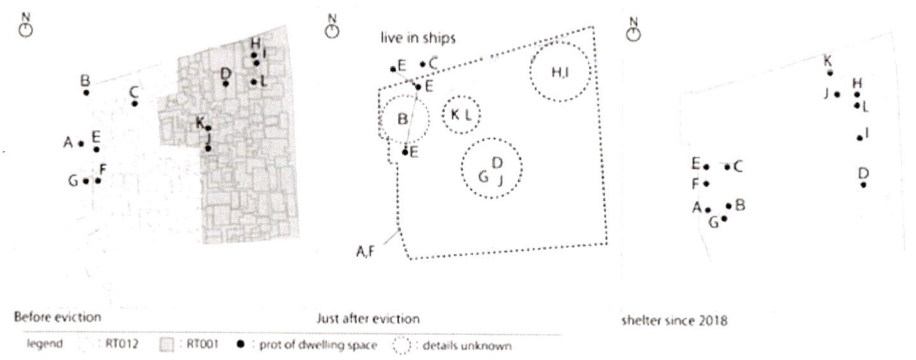

Fig. 4 History of individual's dwelling place

3.2 *Community Based Organizing*

Figure 4 are the plots of individual's dwelling place from before eviction up to the present. This figure shows that it's important for them to live in the almost same place with familiar people. The community leader said that it's one of their history. And we called it "Tempat Nyaman (means Comfortable place)" at that time of meeting. It has been found that people who are familiar with each other actually have a similar Home sphere. Tempat Nyaman is information that is unknown to outsiders, and is an inherent order. In the future, whatever the plan of rebuilding, it will remain as one of invisible order. These facts show that a unique inhabitant culture has been fostered in about 40 years.

Forced eviction has further strengthened the unity of the community. In fact, they are more proactive than ever in improving their own living environment in Kampung. They have tackled the garbage problem, added greenery and gardens, started community businesses, and held numerous meetings. Even in temporary shelters, they have naturally made communal spaces attractive and enjoyable. The question of slum or kampong is clear from the enthusiastic efforts of the Kampung Aquarium, and we can learn a lot from them. These are parts of self-organization by community and are the wisdom of living together.

4 Conclusions

In this paper, we proposed a dialogue method "Sketch-log Dialogue" for understanding and fostering the mechanism of Urban self-organization, especially living environments in neighborhood. The method revealed the Home sphere as perceived and valued by the individuals, the order understood only by the kampung people, and the activities actively undertaken by the community. These facts suggest that self-organization occurs at multiple scales: individual, collective, and community.

In particular, our proposed dialogue method is very useful in understanding the individual scale. In this case, the concept of the home sphere, which is closely connected to personal values, value judgments, life experiences and memories, and can also be grasped as a space, is useful. This is a methodology to learn from people who are actually creating urban landscapes, rather than predetermining the components as an operational concept.

References

1. Allen PM (1997) Cities and regions as self-organizing systems: models of complexity. Gordon and Breach Science Publishers, Amsterdam
2. Batty M (2005) Cities and complexity: understanding cities with cellular automata, agent-based models and fractals. MIT Press, Cambridge
3. Portugali J (2000) Self-organization and the city. Springer, Heidelberg
4. de Bruijn E, Gerrits L (2018) Epistemic communities in urban self-organization: a systematic review and assessment. J Plann Lit 33(3):310–328
5. Eizenberg E (2019) Patterns of self-organization in the context of urban planning. Planning Theory 18(1):40–57
6. Marshall S (2009) Cities, design and evolution. Routledge, London
7. Tunas D, Peresthu A (2010) The self-help housing in Indonesia: the only option for the poor? Habitat Int 34(3):315–322
8. Dian TI (2018) Jakarta's kampungs: their history and contested future. UCLA Theses
9. Elisa S, Amalia NIS, Kiyoko K, Hiroto O (2019) Recovery of evicted kampung after public infrastructure development in north Jakarta—community action planning as innovative people-based solution-1. In: 15th conference of Asian and African city planning (AACP2019). Toyo University, Tokyo, Japan
10. Moroni S, Rauws W, Cozzolino S (2020) Forms of self-organization: urban complexity and planning implications. Environ Plann B: Urban Analytics City Sci 47(2):220–234
11. Ota H, Kanki K (2015) Dialogue for approaching to inhabitant's evaluating sphere in own surroundings in Nishijin, Kyoto: a study on approaching to a wholeness inhabitant's interpretation on city for urban and architectural design. J City Plann Instit Japan 50(3): 1218–1225
12. Bohm D (1996) On dialogue. Routledge Chapman Hall

Urban Fortification and the Contestation of Space: A Case Study in Kayu Putih Housing Complex

Dita Trisnawan, W. Muhamad Rifki Ramadhan, Antony Sihombing, and Achmad Hery Fuad

Abstract Since the tragedy of May 1998 riot in Jakarta, there have been attempts by the community to protect themselves and their properties. The efforts to add protections are achieved through fortification of the area where they live, so that it is closed and difficult to access for outsiders other than the residents. This fortification effort is against the 21st principle of The Charter of New Urbanism that encourages urban areas to be more open, letting cities become friendly places for people, inclusive, and more enliven. This principle aims to create a more dynamic contestation [1] in urban spaces as an effort to improve the livelihood of the city residents, and yet fortification as a protective effort is contradicting the very principle. The Kayu Putih area is chosen as the study area because the author has been part of the residents of the area and witness the development of the fortifications from May 1998 incident up till now, the period of Covid-19 pandemic. Therefore, this study reviews how the fortifications affect the contestation of spaces [2] within the urban setting, and sees its impact on its society [3]. This research uses qualitative method by carrying out field surveys ethnographically in order to access the phenomena and see the impacts that come up. In addition, supporting questionnaires are used to see residents' opinions on fortification, spatial use patterns, and the impact of fortification from time to time on their daily activities and towards the urban space. This study found interesting patterns, especially on the surrounding gates, which became sort of fortresses of the compound/area. Even though the gates closed or have limited access, its existence opens other opportunities for activities, which will not appear if the gates remain open. This research hopefully helps urban designers to consider the expected security [4], as well as the social aspects of fortification and its openness, so that there would be balances between the fortification effort and the contestation of its urban space.

Keywords Contestation of space · Lefebvre · Space production · Spatial representation · Urban fortification

D. Trisnawan (✉) · W. Muhamad Rifki Ramadhan · A. Sihombing · A. H. Fuad
Department of Architecture, Faculty of Engineering, Universitas Indonesia, Depok, Indonesia

L. Yola et al. (eds.), *Sustainable Architecture and Building Environment*,
Lecture Notes in Civil Engineering 161,
https://doi.org/10.1007/978-981-16-2329-5_10

1 Introduction

It cannot be denied that urban development [5] in Indonesia (especially in the city of Jakarta and its satellite area [Jabodetabek]) has become a significant part of physical evolutionary phases that change many non-physical aspects of living, which influence the way people urbanizes.

Fortification as a concept plays an important role in the critical narration of urban development. Fortification is a term that has long become writer's research topic and concern. During these times, many puzzling questions came to mind and these questions could be developed into a deeper research plan for further research writing purpose. Many of these questions were triggered by writer's previous knowledge on fortification within the urban image changes/scope after May 1998 riot in Jakarta (that recorded an architectural appearance post Jakarta's riot). In another scope, further knowledge on Urban Design—specifically about The Lexicon of New Urbanism [6]—delivered The Charter of New Urbanism that contains ideal housing and settlement plan principles. These two early scientific scopes drive the writer to study about fortification term deeper within the urban architecture context, while also try to understand the context in anthropological perspective.

Contestation, according to Merriam Webster dictionary definition means controversy; debate, the word contestation and the word contest also have the same origin, meaning that a contestation can be called a debate, a fight, a bet, a competition in a contest that has a contestant in it. In the context of urban space, urban contestation has contestants in the form of city residents who take part in a contest in order to utilize its urban space. City residents make use of this urban space through various methods, strategies and tactics [3]. The use of urban space within these various perspectives is a form of manifestation of enlivening urban space life [7].

Fortification makes city space [8] into a contest arena that has a certain impacts due to the existence of the fort itself. This study wants to see the relationship between the existence of fortification and the contestation that occurred in the Kayu Putih area. This fortification effort is obviously contrary to the 21st principle of The Charter of New Urbanism that encourages the spirit of openness in urban space, since this principle expects more open spaces for interaction between urban residents. This study wants to examine the presence of fortification, which is an effort to provide a sense of security from threats with the level of contestation that occurs, especially by considering the context of the Covid-19 outbreak conditions, that eventually have encourage the increased of fortification efforts.

2 Methods

2.1 *Study Area*

The area as the focus of the research is the residential area in Kayu Putih Sub-district. This area is bordered by Jalan Tener on the northern side, Jalan Sirap on the eastern side, Jalan Angkur on the southern side, and Jalan Pondasi and Jalan Pondasi Ujung on the western side (Fig. 1). The study uses a housing complex in Kayu Putih, since this area has a dynamic chronology of fortification process, recorded by the residents and the author in mind.

At first, this area was an area that was free to access by anyone. Then, the tragedy of the May 1998 riots in Jakarta occurred and had an impact on the sense of security of the residents living in it, so they made a barrier (as a fort) to protect them from the threats that existed at that time. Along the way, after the tragedy (incident) passed, the fort that had been formed was maintained and even strengthened because threats remained, such as the threat of criminal acts, like threats of theft, pick pocketing, kidnapping, and others.

However, with another incident currently happening, the Covid-19 pandemic, the existing fortifications need to adapt and the way of adapting to the situation makes it necessary to delve deeper into the phenomena that occur. This situation re-tests the contestation that has occurred and will change the way the designer's point of view in designing and responding to the possibilities that might exist in the future.

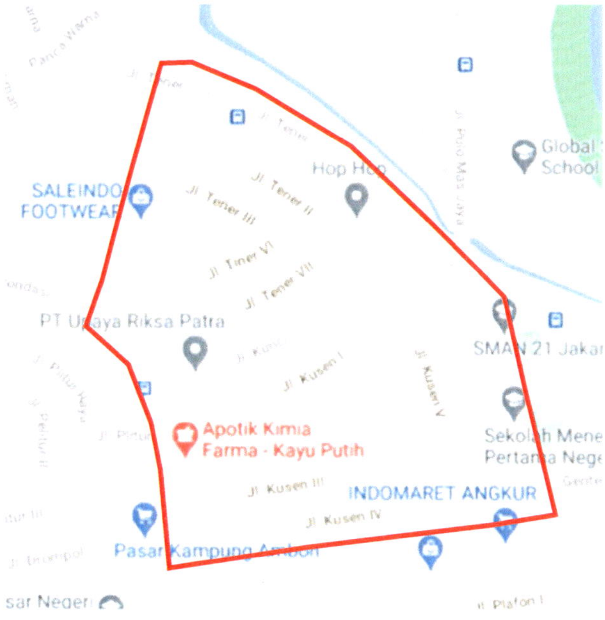

Fig. 1 Location map of study area

2.2 *Data Processing*

Understanding urban fortification topic in big cities and uncontrolled slum area context can be achieved by ethnographic study [9]. Therefore, it can be investigated how these differences are negotiated and presented. Watson revealed how borders between public and private can be negotiated and redrawn, and how public and private spaces laid out each other. According to this research, author considers the fact on how urban environment is affecting someone's decision to break the rule. The analysis result, therefore, is used to find answers to questions that come from the city, especially in social interaction within urban spaces (where usually guided by various land use functions and people activities).

In seeing and reading the phenomena that occur in housing complex of Kayu Putih area, the author observed using the ethnographic method so that the possibilities and findings appear authentically. This method will also allow the reader to experience the flow of this study by providing an understanding of the phenomena that occur.

The reading of the phenomenon will be delivered by carrying out a field survey (ethnographically) to find patterns of society [10] in using fortified urban spaces and seeing its excesses (which can be seen as a strategy in city) [11]. Apart from field survey, this study also uses a questionnaire for residents of the complex containing maps so that respondents can show the ways they use urban space. In addition to the map, the supporting questionnaire which contains questions that support the map, seeing respondent opinions about fortification condition in the area, as well as exploring its community activities as part of the contestation within the area.

The knowledge gained about fortification was then linked to the patterns [12] and findings about contestation that occurred in this area. Then, this linkage is analyzed by reviewing The Charter of New Urbanism [6], especially on the 21st principle. The final goal, the author hopes that policy makers, urban designers, and architects in general can use the study as an alternative consideration, as the urban principles in The Charter of New Urbanism are not always applicable (even contradictory) to its current conditions and open the chance of renegotiating the principles to each specific context.

3 Results and Discussion

3.1 *Contestation of City Spaces in the Kayu Putih Area*

The Kayu Putih area of the research study has five gates that affect the surrounding area and into the inner side of its housing area. These five gates affect the surrounding area according to the context of their respective locations. These gates are then called Gate A, B, C, D and E (Fig. 2). Gate A is the main gate that becomes access for all modes of transportation, both pedestrians, car and motorbike riders,

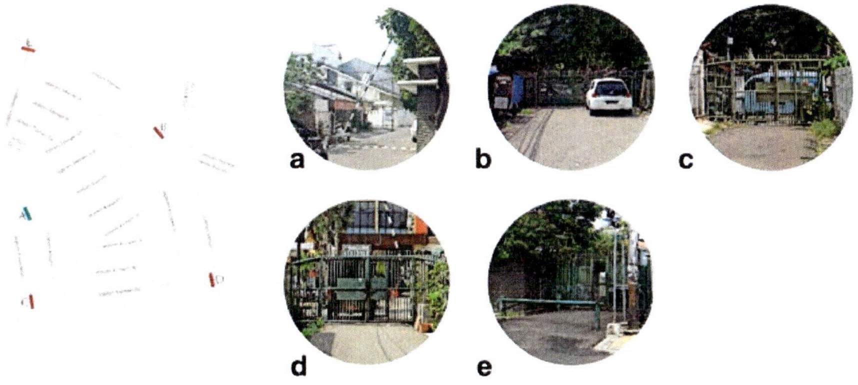

Fig. 2 The five gates which are the access to the housing complex

and cyclists. Gate A is in the form of a portal crossbar guarded by a security unit equipped with a guard post and closed circuit television (CCTV) surveillance. The other four gates, Gates B, C, D and E are secondary gates with iron gates, approximately 2–2.5 m high with a 30 cm barrier at the bottom so that only pedestrians and cyclists can pass through them (note that to pass through these gates they need to lift the bicycle).

After the Covid-19 pandemic, the gate system in this complex underwent changes. The four secondary gates are permanently closed, leaving only Gate A which is the main gate as the only access in and out of the complex. The closure of these four secondary gates is a form of response to maintain the security and safety of residents from the outbreak.

After the Covid-19 pandemic, the gate system in this complex underwent changes. The four secondary gates are permanently closed, leaving only Gate A which is the main gate as the only access in and out of the complex. The closure of these four secondary gates is a form of response to maintain the security and safety of residents from the outbreak because everyone who goes in and out becomes more monitored and the potential for virus entry is expected to be more controllable. This stricter access restriction further reinforces the contradiction to the 21st principle of The Charter of New Urbanism which encourages urban spaces to become more open and accessible. This tightening of access is not only contradictory, but is something that must be done. The aspects of human safety are more important, because essentially the need for security is more fundamental than the social need (Maslow 1943). This very need goes against the 21st principle of The Charter of New Urbanism. This tightening of access will affect the dynamics of contestation that occur as a form of response and changes in community strategies in utilizing the urban space [13].

The Fig. 3 graph shows the intensity of gate use before the Covid-19 pandemic. The data shows that Gate A (the main gate) is the most often passed gate before the tightening in the condition of the Covid-19 pandemic. The other gates were still being used, where Gate C in particular was quite intensely used, being the second

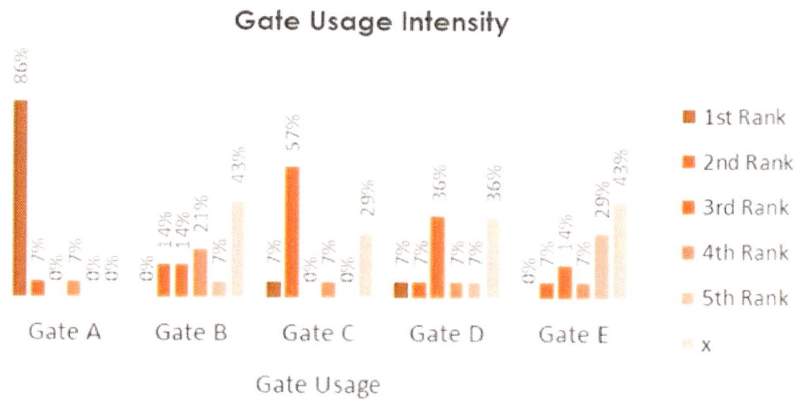

Fig. 3 Gate usage intensity before the Covid-19 pandemic

highest priority in its use. At each gate (as a form of fortification), there was a different contestation and the intensity on the use of the gate. It had an impact on the contestation that occurred in the surrounding area and within the housing environment.

As the main gate, Gate A gets abundant volume of pedestrian traffic (Fig. 4). This happens because people who usually use Gates B, C, D, and E as their main access forcefully use this gate. The residents still willingly do this even though their homes are far from the gate. For motorized vehicle users, the traffic volume remains the same, since before the Covid-19 outbreak all vehicles have indeed used the Gate A as a result of closing other gates for motorized vehicle access.

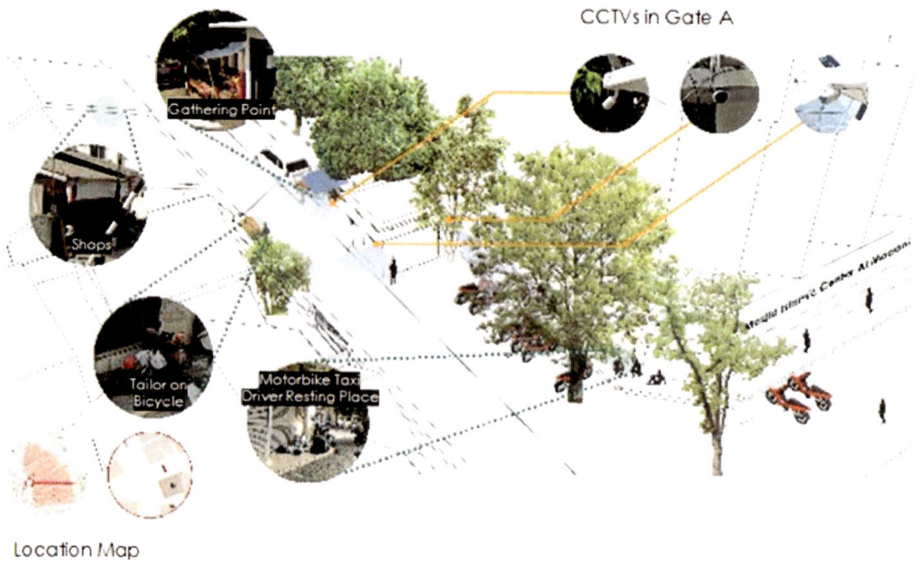

Fig. 4 Contestation in gate A surrounding area

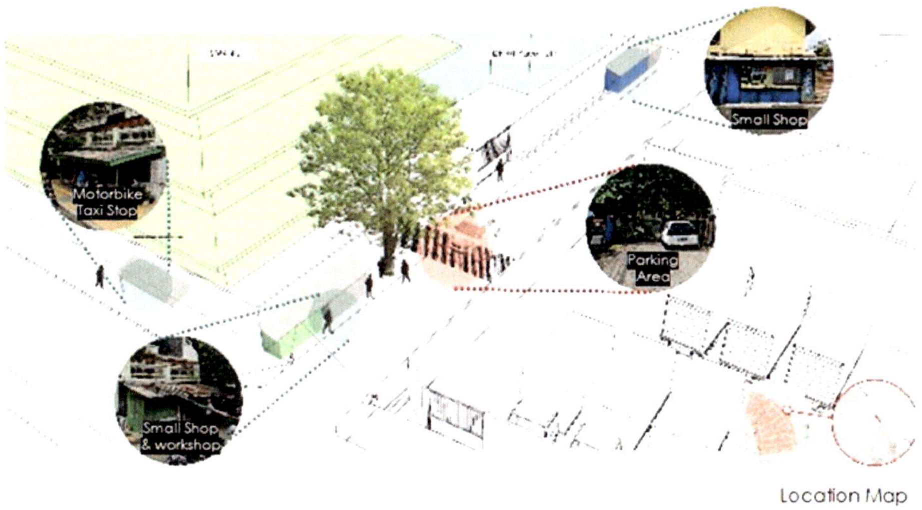

Fig. 5 Contestation in gate B surrounding area

The condition of Gate B before the Covid-19 pandemic was quite busy with two schools, SMA Negeri 21 Jakarta (Public High School) and SD Negeri 01 Kayu Putih (Public Elementary School). The existence of SDN 01 Kayu Putih has a direct effect on the conditions around Gate B because the school's entrance is in the inner side of the gate, while SMA 21 has no direct effect because the entrance is in the outer side of the gate. The existence of this elementary school brings up small stalls where students can snack on, also for people dropping off or picking up the students. Pandemic condition requires this school to cancel most learning activities at school, thereby drastically reducing human traffic, coupled with the closure of Gate B, affecting stalls to also close their business activities. Apart from stalls stopping their business, the presence of motorcycle taxi drivers, just outside the gate, has also disappeared (Fig. 5).

In the Gate C area, there are shophouses facing two sides of the road, Jalan Kusen IV and Jalan Pondasi. The closing of the gate made the shophouses facing the side of Jalan Kusen IV closed, since they were not visible any longer from the main road (Jalan Pondasi) and no longer passed by people because of the gate's closure. This did not happen to the mosque shops close to the Gate A, since the mosque decided to move back the gate as far as the shophouses stretched, so that the shop area can be accessed freely by everyone. Because of the position of Gate C is right across the Kampung Ambon Market and right next to a bank office, sellers use the area to sell chicken noodles (in the morning) and *pempek model* (in the evening). *Angkot* (minibus public transportation) drivers also use the area as a place for them to stop to find passengers (*ngetem*). However, due to pandemic situation, the closing of Gate C has the potential to make it more difficult for residents who want to buy food from stalls or want to use public transportation, since they have no direct access. Fortified gate discourages them to the outsides, since accessing it has become more difficult (Fig. 6).

The position of Gate D, which is right across the Kayu Putih Sub-district Municipal Office, let the outer space used by the officer to park *Satpol PP* (local police) car. In addition, on the inner side of the gate there is a small shop that since the pandemic condition, losing the access to potential shoppers. This potential lost happens because residents do not pass through Gate D any longer, not passing the shop, and make the potential for incidental shopping is lost (Fig. 7).

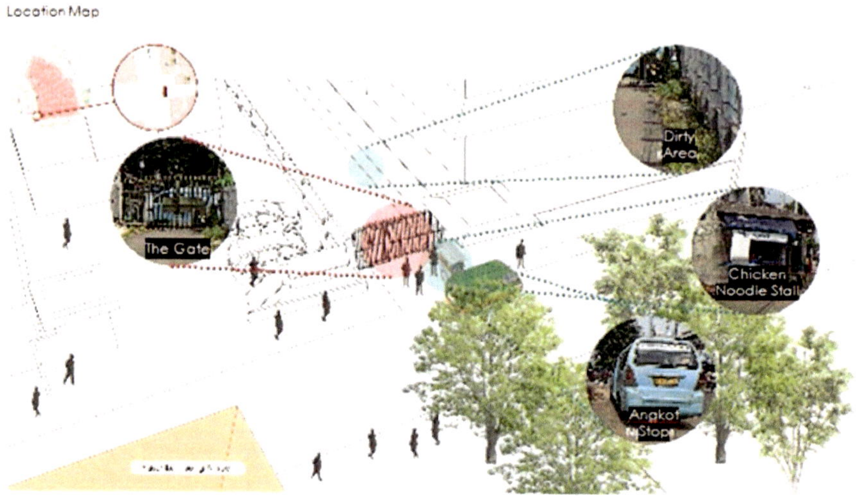

Fig. 6 Contestation in gate C surrounding area

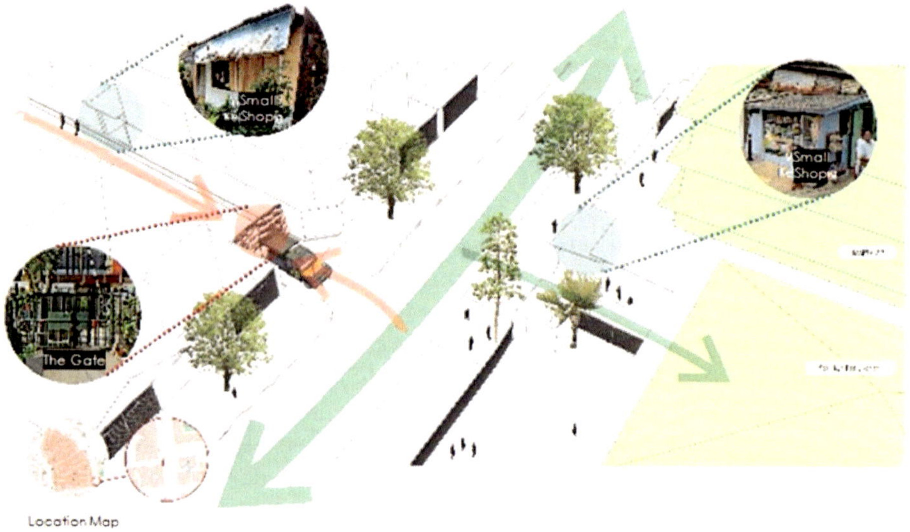

Fig. 7 Contestation in gate D surrounding area

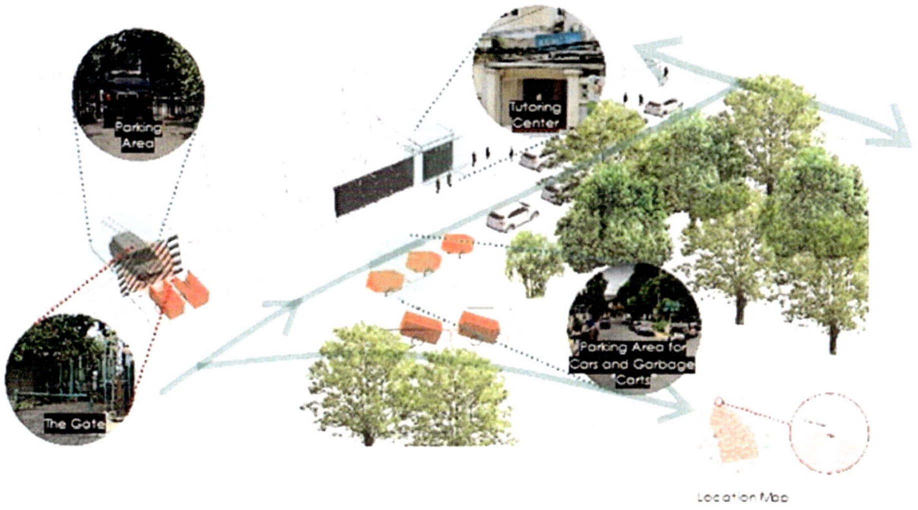

Fig. 8 Contestation in gate E surrounding area

The outer side area of Gate E, which is not crowded with people, has made garbage men use the area as a place to park their garbage carts (Fig. 8). In the area outside Gate E, there is also a tutoring center. Before the pandemic, tutoring students also used the strip to park their private vehicles. The existence of these tutoring students also brings up hawker vendors around the area. Meanwhile, on the inner side, residents whose houses are near the gate take advantage of the area as a parking area for their automobiles.

3.2 *The Linkage of Fortification and Contestation of City Spaces*

With the presence of fortification in this area, the community took advantage of the areas that were formed as a result of the fortification (with the existence of the gates). These areas include the area around each gate as well as the area within the fortified complex. From each gate that was observed, the presence of fortification brought a pattern of activities that became the public's response to the existence of the gate. Abundance traffic through Gate A, has made it strategic area for trading. Situation has then made the occurrence of mobile hawkers such as tailorman on bike and instant drinks seller. On the other hand, the Gate E area, which has quiet traffic coupled with fortification, has made the area even quieter and is used to park garbage carts, something that is not going to happen in areas that are easily seen and close to areas of activities.

Figures 9 and 10 show how the residents feel about the condition of the area around the gates, both the outer side and the inner side. These graphs illustrate how the contestation that occurs becomes an opportunity for its residents. For example, some residents use the inner area for parking lot because the area becomes quiet after the gate was built. Local merchants also take advantage of the busy outside area to have a small business. Contestation that occurs due to the impact of fortification is not a bad thing for residents, even the survey results show that 71% do not feel constrained by the existence of gates in their residence. These gates provide them with a sense of security (including by changing the gate system during the Covid-19 pandemic) and are also used to support the livelihood of the residents.

Besides having an impact on the area around the gate, the presence of fortification has an impact on the residential environment as shown in Fig. 11. In this diagram, it can be seen that the inner area close to the gate and the axis of Gates A-B has a worse condition than other areas. This occurs as a result of a quiet area because it is not traversed by people and vice versa, there are areas that are very crowded with people, such as the area along the axis of Gates A–B.

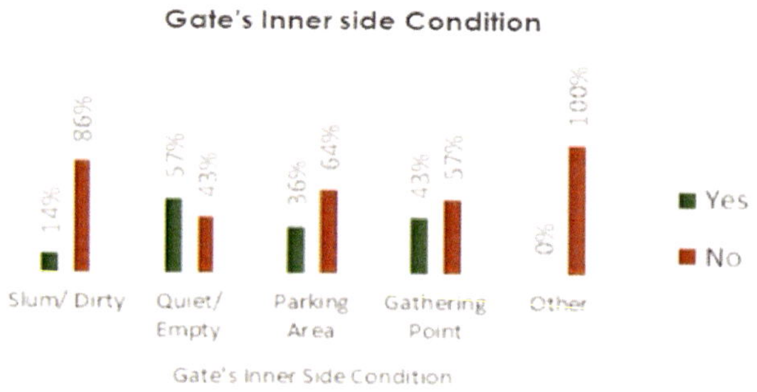

Fig. 9 Graphic of gate's inner side condition

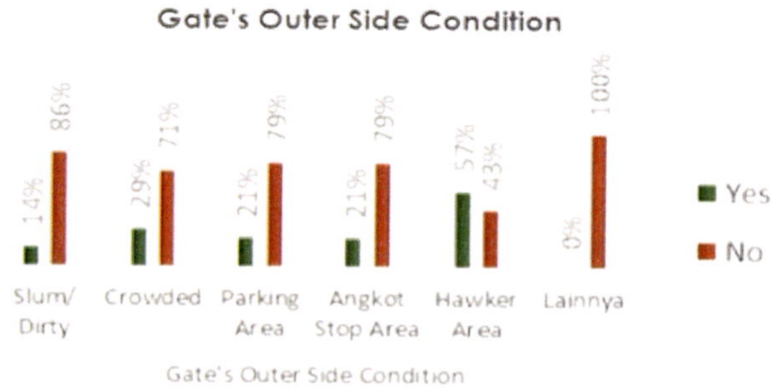

Fig. 10 Graphic of gate's outer side condition

Fig. 11 Diagram of Kayu Putih's housing condition

4 Conclusions

The presence of fortification in the Kayu Putih area in the form of gating the residential area shows the emergence of contestation which is the result of adaptation to the prevailing conditions. This adaptation process can be taken into consideration when designing an area or city, because fortification that hinders openness does not always hold contestation to a city. The incidents that occurred since the May 1998 riots until the Covid-19 outbreak showed that residents were able to interpret the use of city space [11]. Citizens as parties who play a role in carrying out city life [14] must be the main focus beyond the design made by the designers. Strategy and tactics will always appear in forms that the designers had not thought of. City residents are the ones who always better in understanding their own city space. In designing a new area, instead of making design that are forcing and dictating the movement patterns of the people, it would be wiser for the designers to provide space for the growth of public creativity so that the contestations that occur in urban spaces are even livelier.

Acknowledgements This paper was prepared authors and funded by Universitas Indonesia through PUTI 2020 Research Grant.

References

1. Contestation (2011) In: Merriam-Webster.com. Retrieved 20 Oct 2020, from https://www.merriam-webster.com/dictionary/contestation
2. Featherstone D, Korf B (2012) Introduction: space, contestation and the political. Geoforum 43:663–668 (Elsevier)

3. Next Generation of New Urbanist (2011) Tactical urbanism
4. Ceccato V (2012) The urban fabric of crime and fear. Springer, Dordrecht
5. Hansen NM (Apr 1973) Urbanism and regional disparities in developed countries. Rev Soc Econ 31(1):54–65, Taylor & Francis. Stable URL: http://www.jstor.org/stable/29768677. Accessed on 29 Apr 2014 (03:37)
6. Duany A, Plater-Zybrek E (1999) The Lexicon of the new urbanism. DPZ, Miami
7. Wirth L (Jul 1938) Urbanism as a way of life. Am J Sociol 44(1):1–24, The University of Chicago Press. Stable URL: http://links.jstor.org/sici?sici=0002-9602%28193807%2944%3A1%3C1%3AUAAWOL%3E2.0.CO%3B2-B
8. Dikeç M (2012) Space as a mode of political thinking. Geoforum 43:669–676 (Elsevier)
9. Watson S (2006) City publics: the (dis)enchantments of urban encounters: questioning cities
10. O'Connor RA (Mar 1995) Indigenous urbanism: class, city and society in southeast Asia. J SE Asian Stud 26(1):30–45, Perspectives on Southeast Asian Studies. Cambridge University Press on behalf of Department of History, National University of Singapore. Stable URL: http://www.jstor.org/stable/20071699. Accessed on 22 May 2014 (04:21)
11. Pløger J (2008) Foucault's Dispositif and the city, Roskilde University, Denmark, plo@ruc.dk
12. Clark D (Jan 2000) World urban development: processes and patterns at the end of the twentieth century. Geography 85(1):15–23, Geographical Association. Stable URL: http://www.jstor.org/stable/40573371, Accessed on 29 Apr 2014 05:38
13. Borden I (2000) The unknown city, contesting architecture and social space. Massachusetts Institute of Technology, Massachusetts
14. Karp DA, Stone GP, Yoels WC (Summer 1992) Being urban: a sociology of city life, symbolic interaction, 15(2):245–246, Review by: Reitzes, DC, Wiley on behalf of the Society for the Study of Symbolic Interaction, Stable URL: http://www.jstor.org/stable/10.1525/si.1992.15.2.245. Accessed on 05 June 2014 (00:18)

Numerical Study on the Effect of Single and Multi-injection Pressurized Fan in Enclosed Stairwell of High-Rise Building

Beline Alianto, Nasruddin, and Yulianto Sulistyo Nugroho

Abstract This study aimed to evaluate the pressure difference distribution across the escape stair in a 24-story stairwell. A large eddy simulation (LES) method was used to model the pressure in a stairwell model with a set of simulation parameters using FDS (Fire Dynamic Simulator) software. Two configurations of fans involved single injection, and multiple fan injection is observed. A Single-injection fan is located on the highest point of the stair. The double injection system consists of two fans and is separated at 45 m at each other. When the fire alarm is activated, the pressurization fan automatically runs and delivered outdoor air to the stairwell. It is commonly acknowledged that the pressure would start to built-up in the stairwell before the occupants entering the stair by open the door. Besides, when the pressure is significantly high, the door is unable to be open from outside by occupants as a consequence. In order to overcome this issue, a pressure relief damper was then installed on the top of the stair. The numerical study shows possible pressure difference in the stairwell is about 50 and 95 Pa. However, a multi injection system has the advantages of proper pressure distribution along the stairwell. Performance-based design by using numerical computation is a valuable tool to predict the engineering system and presents a better understanding of how the system would perform in design.

Keywords Stairwell · Pressurization · Single-injection · Multi-injection · Pressure difference

B. Alianto · Nasruddin · Y. S. Nugroho (✉)
Department of Mechanical Engineering, Faculty of Engineering, Universitas Indonesia, Kampus UI Depok, Depok 16424, Indonesia
e-mail: yulianto.nugroho@ui.ac.id

L. Yola et al. (eds.), *Sustainable Architecture and Building Environment*, Lecture Notes in Civil Engineering 161,
https://doi.org/10.1007/978-981-16-2329-5_10

1 Introduction

The stairwell is a primary means of escape in a high-rise building. During the fire incident, occupants would exit the building by the stairwell. During high-rise fires, smoke can be contained and controlled in compartmented with stairwell pressurization systems. Smoke is inhibited from entering the stairwell.

A preliminary study of air leakage was done by Tamura [1]. Thomas [2] also conducted experiments using a wind tunnel 90 cm square on the movement of hot gases and smoke revealed that minimum air velocity is necessary to prevent the smoke from flowing back. The mathematical equation of calculating the pressure difference of the compartment was introduced.

A staircase cannot be both pressurized and opened to outside at the same time [3]. Jensen suggested that about 20 Pa minimum pressure is required in the stair. The average pressure inside the stairwell is about 2 $Pa.m^{-1}$, the height of the stairwell [4]. Stairwell pressurization is considered not-effective when the stairwell door is open [5]. In this case, when the air velocity at the open stairwell door was greater than 1.0 m/s, the wind speed method is an appropriate method to calculate the mechanical smoke control volume flow rate of the pressurization system [6]. The study aims to provide a simple description of single and double injection stairwell pressurization systems so as to resolve the research challenges. This can help identify actual performance-based of the applied design engineering and aid in designing a clearer safety design model.

2 Basic Theory

The method of pressurization system can be classified into single injection and multiple injections. A single injection system has a pressurization fan at one location, which is commonly on the top of the stair. A single injection system might fail if a stairwell access door opens near the injection point. Loss of pressurization air will occur immediately at the location which is farther from the injection point. The multiple injection system has sources of supplied air from more than one location by using multiple fans or a single fan connected to a ducted arrangement with multiple outlets. The air is delivered to every floor of the stairwell by means of a pressurization shaft. The stairwell is commonly provided with a smoke-free lobby (Figs. 1 and 2).

3 Modelling Setup

The simulation used a 24-story stairwell, which is 92 m high, as depicted in Fig. 3. The floor to floor is 4 m high. The cross-section of the stairwell is 6.4 m × 3.6 m. There is a smoke-free lobby on each floor between the exit of the stairwell and

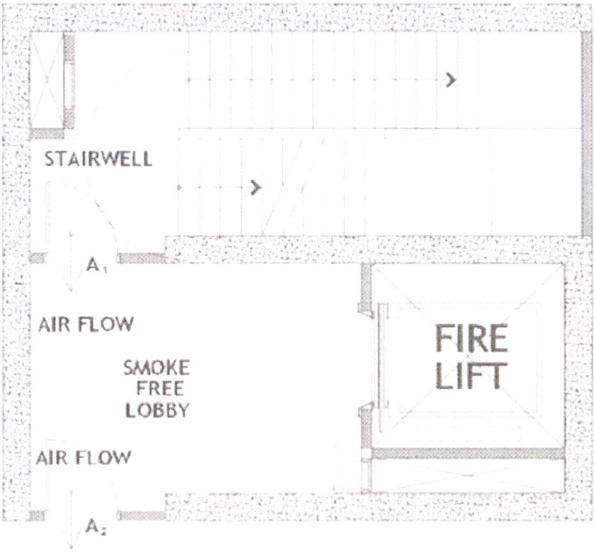

Fig. 1 Stairwell and smoke free lobby

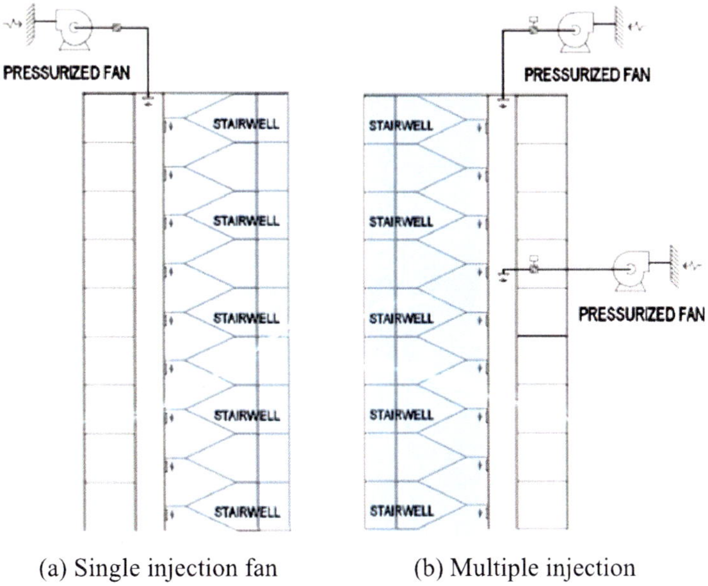

Fig. 2 Single injection and multiple injection stairwell pressurization

adjacent compartment, which creates a series of airflow from the stairwell to the lobby and compartment on each floor. The dimension room size of the smoke-free lobby is 3.6 m × 2.2 m. One escape door is connected between the stairwell and

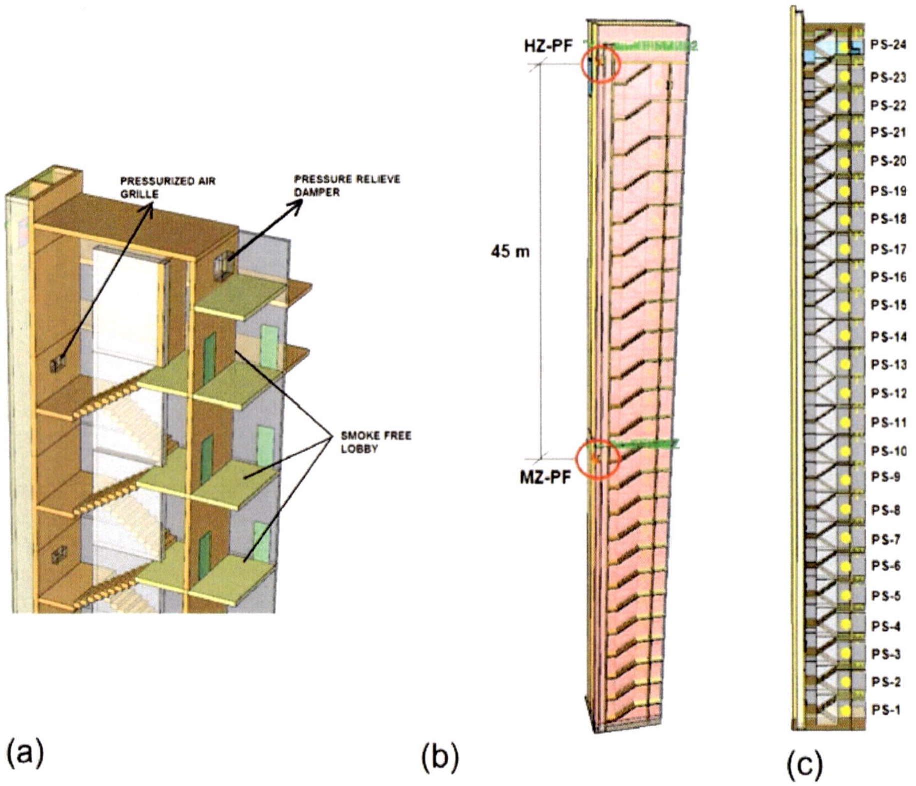

Fig. 3 Geometry of stairwell included for pressure air grille, pressure relieve damper, and smoke-free lobby

lobby, and another is connected smoke lobby and compartment on every floor. Pressurized air is injected through a vertical shaft that connected every floor to the stairwell via a pressurized grille. The pressurized air grille opening in the stairwell is 0.8 m × 0.4 m. Grilles are provided on every two floors. A pressure sensor is located at every floor inside the stairwell to measure the pressure difference between stairs relative to the outside. The Single-injection scenario has only one pressurized fan on the top of the stair shaft. The double injection system has two sources of pressurization air by a fan on the top of the stair, and another is in the middle height of the stair. The distance between fans is set 45 m.

The full-scale stairwell's internal lining is specified as "CONCRETE", and its density, conductivity, and specific heat are 2200 kg/m^3, 1.2 W/(m K), and 0.88 kJ/(kg K). The arrangements of measuring points are located in the stairwell and smoke-free lobby. In the study, the door is controlled to be open after 90 s the fire is ignited. The size of the compartment is 12 m × 6 m. The door is open on the affected fire floor to be evacuated so that more air can be supplied through this door to inhibit smoke spread into the stair.

According to the standard practice and norm, the pressure difference across the stairwell should be maintained at 37.5–82.5 Pa. It is mostly required the pressure difference at 50 Pa as required by many prescriptive codes. The purpose is to prevent the occupant unable to open the door from outside during a fire alarm is initiated. A damper is modeled on the top of the stair, which is opened when the pressure rises above 82.5 Pa and closed when the pressure dropped below 37.5 Pa. The size of the damper is 800 mm × 600 mm. Figure 2 shows the simulation model.

4 Numerical Solution

Each simulation was repeated under scenarios:

- Single-injection fan q = 5 m^3/s
- Double injection fan q = 2 × 2.5 m^3/s

In this simulation, the doors of the stairwell are considered closed at all the time. During the activation of an alarm, a pressurization fan would automatically run and supplied pressurized air into the stairwell. Normally, occupants would require a period of time to verify the incident. If the fire confirmation is true, then the occupants would decide to evacuate and exit the building through the stairwell. Soon after the fan is running, the pressure starts to build-up in the stairwell. When the pressure is noticeably high, occupants are not able to open the door from outside. On the contrary, excessive low pressure would allow smoke to enter the stair.

In the beginning, the pressure is increased steadily. The air is injected from the top of the stair. The fan is running at 5 m^3/s in a single injection system. The pressure reading from the pressure transducer is plotted on the graph. The simulation time is taken 300 s. The fan flow rate of 5 m^3/s produced a slightly lower pressure range at 50–95 Pa for the highest and lowest floor. In this case, the stairwell pressure produced by fan flow rate at 5 m^3/s is fair in terms of the allowable door opening force by occupants. During the early stage of an operation, the pressure built up at the lower zone, and then finally, the upper level has higher pressure. Soon after, the maximum pressure moved toward the lower level and repeated until the pressure balance is achieved.

The pressure difference over time curve for both single and multiple injection can be seen in Fig. 4. The effect of the pressure profile of a multi-injection pressurized system is observed. Two fans (2 × 2.5 m^3/s) are used instead of a single fan. One fan is supplied outdoor air from the top of the stairwell, and the other is on the intermediate floor at level 12. The possible pressure difference in the stairwell is about 50 and 95 Pa. To conclude, in terms of pressure balance across the stairwell, multiple injections can be seen as a better method.

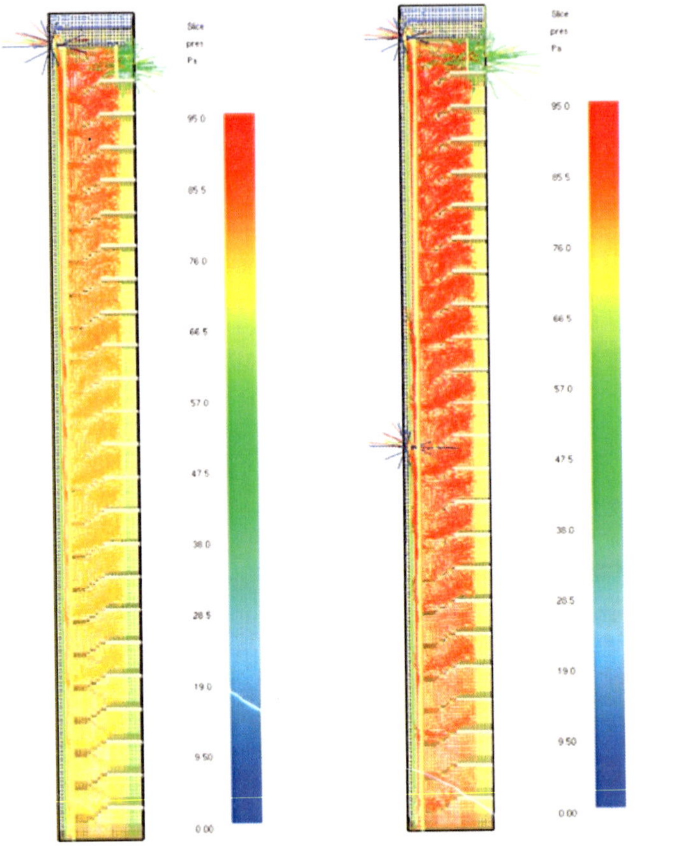

(a) Single Injection
$q = 5\ \mathrm{m^3/s}$ (t = 113 s)

(b) Double Injection
$q = 2 \times 2{,}5\ \mathrm{m^3/s}$. (t = 250 s)

Fig. 4 Pressure profile for single and double injection pressure fan

5 Conclusions

In many buildings, stairwells are mostly provided with a single injection pressurized system. A single injection system might cause unbalance pressure distribution along the stairwell. Multiple injections are preferable in terms of maintaining balance pressure. Prescriptive codes offer the solution of designing the stairwell pressurization through the design criteria standard set for allowable pressure difference. An analytical solution is common used by an engineer. However, the prediction of the solution given in both methods does not always present the performance approach in many cases. Because of the irreversible process during construction, unpredicted issues by analytic method do not always cater to the design. Therefore, computer numerical simulation is not only a cheaper solution but also as a tool for virtual building mock-up to predict the real phenomenon of likelihood to happen before the design is applied.

Acknowledgements The authors would like to thank the financial support provided by the Ministry of Research, Technology, and Higher Education of the Republic of Indonesia through Penelitian Terapan Unggulan Perguruan Tinggi (PTUPT) 2019–2020 funding scheme managed by the Directorate for Research and Public Services (DRPM) Universitas Indonesia.

References

1. Tamura GT, Wilson AG (1964) Air leakage measurements on two occupied houses. ASHRAE Trans 70:110–119
2. Thomas PH (1970) Movement of smoke in horizontal corridors against an airflow. Inst Fire Eng Q 30(77):45–53
3. Hansen ND, Stevenson FB, Valkist M et al (2018) Movement of smoke in horizontal corridors against an airflow. Inst Fire Eng Q 30(77):45–53
4. Jensen L (2003) Stairwell flow pressurization—a new method. J Fire Prot Eng 13(4):251–274
5. Klote JH (2004) Tenability and open doors in pressurized stairwells. ASHRAE Trans 110 (7):613–637
6. Chen YL, Shicheng DW (2010) Application analysis of wind speed method in mechanical smoke control volume flow rate calculation. Fire Sci Technol 29(11):958–961

Public Service Agency for Institutional Form of Social Housing in Jakarta

Rini Kurniawati, Joko Adianto, and Lita Sari Barus

Abstract This paper aims to understand the substantial reason for the surmounting arrears in GORA (government-owned rental apartment) and hypothetically conclude a newly-proposed institutional form to overcome the issue. GORA is an appropriate type of house for assisting the low-incomes to meet their housing need in Jakarta. Although the rental fee is considerably affordable and subsidized by the municipal government, the arrears are surprisingly high. This condition is burdening the municipal budget expenditure for maintenance and operational costs, which leads to degrading building conditions. Through the systematic literature review method, this paper examines the GORA affordability for the low-income tenants and the appropriate institutional form for GORA management for overcoming the arrears. The finding shows most of the tenants have potential failures to meet the affordable rental fee due to their low monthly income. Therefore, the Public Service Agency (PSA) becomes the newly-proposed institutional form for GORA management, as it is welfare-oriented with flexible financial management which allows it to generate income for subsidizing the maintenance and operational costs without increasing the rental fees.

Keywords Government-owned rental apartment · Affordability · The low-incomes · Public service agency · Jakarta

R. Kurniawati · L. S. Barus
Urban Development Study, School of Strategic and Global Studies, Universitas Indonesia, Depok, Indonesia
e-mail: rini.kurniawati91@ui.ac.id

L. S. Barus
e-mail: litabarus@ui.ac.id

J. Adianto (✉)
Department of Architecture, Faculty of Engineering, Universitas Indonesia, Depok, Indonesia
e-mail: joko.adianto@ui.ac.id

L. Yola et al. (eds.), *Sustainable Architecture and Building Environment*, Lecture Notes in Civil Engineering 161,
https://doi.org/10.1007/978-981-16-2329-5_13

1 Introduction

DKI Jakarta, the capital city of Indonesia, has been experiencing massive urbanization. The total population of this province in 2017 approximately reaches 10.5 million with an average growth of 1–2% per year and indicates a severe housing shortage in the capital city [1]. In 2018, the housing backlog in the capital city is approximately 300,000 and escalating in the future if the housing supply fails to cope with the housing demand [2].

This condition, based on the basic economic principle, increases the housing price. According to the Ministry of Public Works and Housing (MPWH) decree 242/2020, the housing price in the capital city reaches IDR 315–330 million with 21–36 m^2 house size, for those who earn a monthly income less than IDR 8 million. While there are 3.77% of population in 2017 earn a monthly income less than IDR 600,000 [3]. This population group is surely excluded from the affordable housing program, due to their poor financial capacity.

Social housing, or Government-Owned Rental Apartment (GORA in Indonesian), is a preferred solution to meet the excessive housing demand with the limited size of the available land. It is considered as a suitable housing type for the low-income, who cannot afford to purchase a house in the capitalistic housing market, to live in the subsidized rental house with an affordable rental price.

GORA poses latent yet grievous problems. One of the observable problems is the surmounting arrears of low-income renters. The amount of arrears reaches more than IDR 72 billion, from approximately less than 70% of total units [4]. The staggering numbers of arrears are potentially increasing with the enacted Governor of DKI Jakarta decree 61/2020, which defers the obligation of rental payment to GORA's renters in pandemic times.

This paper is a critical literature review of the institutional reform of GORA management and formulates a hypothetical answer to this dilemmatic problem. It begins with a critical discussion on the literature review of social housing to identify the institutional problem according to worldwide experience. It is followed by the analytical review of the Indonesian experience in the management of GORA. This series of reviews provides a solid ground for a newly-proposed alternative institutional form to manage GORA.

2 Social Housing and Its Management: A Worldwide Experience

Essentially, social housing should not be considered only as of the shelter provision for the low-incomes but also welfare improvement. Therefore, social housing should provide long-term accommodation for income-weak households to ensure the fulfillment of their basic need, as income improvement is rarely achieved in the short-term period. Furthermore, a study emphasizes the welfare-oriented, not profit-oriented, management to ensure the low-income households meet their shelter need [5].

The idea for hybridizing the public and private funding enters the conversation of social housing in Europe since the second millennium to support the professional management and sufficient funding for subsidizing the low-income tenants [6]. A recent study suggests the utilization of the under-utilized municipal land for social housing, rather than purchasing land with an unaffordable market price [7]. The enactment of housing authority enables the self-sufficient management of social housing [8]. Therefore, a new type of management system and body is required to provide self-sufficient funding for subsidizing maintenance and operational costs of GORA without burdening the low-income tenants with a high rental fee.

Is GORA affordable for low-income households in the Province of DKI Jakarta? Then, with the surmounting arrears, what kinds of an appropriate newly-proposed institutional form for self-sufficient GORA management?

3 Methodology

This paper utilizes the literature review as the methodology for uncovering areas, building conceptual models, and providing theoretical frameworks in future studies through synthesizing findings on a meta-level [9]. It takes qualitative systematic reviews as a process to critically appraise the relevant prior studies for formulating the profound hypothesis [10]. This type of review has been praised for minimizing bias, as it utilizes explicit methods to review the collected secondary data for the conclusion. The collected data for answering the first question such as various historical literature, national and provincial regulations relating to GORA, also the recent demographic statistic of Jakarta from CBS. The current data of GORA's arrears in Jakarta is also gathered from PHSA DKI Jakarta. They are analyzed descriptively to understand the probability of GORA affordability with the average monthly income and expenditure of the low-income tenants. While for the second question, recent publications of GORA's management problems are collected and analyzed to obtain a profound institutional issue. Besides that, In-depth interviews are conducted with GORA management to verify the collected secondary data and analysis results.

4 Discussion

4.1 The Affordability of GORA

The income majority or 54% of the DKI Jakarta GORA residents is from IDR 1,000,000 to 4,000,000 per month; 32% of them have an income below 1,000,000

IDR per month, and 14% of them have an income above IDR 4,000,000 per month. It can be seen that the majority of GORA residents in DKI Jakarta, or 86% of them, have an income below the regional minimum wage of DKI Jakarta in 2020 (IDR 4,276,349), and only 14% of them have an income above the regional minimum wage of DKI Jakarta [11].

The high rate of arrears in the GORA of DKI Jakarta illustrates the residents' household condition's inability to allocate a part of their household income for the housing cost, in this case, the rental rate of GORA. In the Regulation of the Governor of DKI Jakarta Number 55 of 2018, it is explained that the average rental rate for the block-type GORA, namely the programmed community rate, is IDR 300,000, and for the general community is IDR 600,000; while the high-level (tower-type), namely the programmed community rate, is IDR 400,000, and for the general community is IDR 700,000. If we examine the data on the residents' income amount living in the GORA of DKI Jakarta, which is that 32% of them have an income of <IDR 1,000,000 per month, so with the lowest rental rates of IDR 300,000 and IDR 400,000 per month (the rates of the programmed community), as many as 32% of them resulting in housing affordability problems. This is because they have to spend more than 30 percent of their household income for the monthly rental rate, and what happens is that they are unable to pay the housing cost, in this case, the rental rate, thus resulting in the rental arrears.

Therefore what kind of a newly proposed housing authority or management body enables the self-sufficient financing to subsidize the affordable rental fee for the low-income tenants in Indonesia, in particular Jakarta?

4.2 *The Urgency for Revising the Management System*

4.2.1 Municipal Government as GORA Management

The GORA management by the municipal government serves relatively poor results, not only in the capital city but also nationwide. A study in Kemayoran GORA (Central Jakarta) shows that poor maintenance for GORA management lowers the housing satisfaction of the low-income tenants and implicates the unwillingness-to-pay the rent fee [12]. It is worsened by the surmounting numbers of arrears that leads to deteriorating GORA physical quality. Consequently, the physical building deteriorates and endangers the safety of the low-income tenants [13]. The management of GORA by the municipal government receives low satisfaction from the low-income tenants because of the limited and inflexible cost for maintenance, which leads to degraded housing quality [14]. Based on data from Public Housing and Settlement Agency (PHSA) DKI Jakarta, 24,916 units of GORA currently, around 18,444 units are occupied, and while the other 6472 units are unoccupied yet due to physical damages [4]. These numerous damaged units imply a lack of funds for repair.

In terms of GORA, Apartment Law 20/2011, and the State Ministry of Public Housing decree 14/2007 Paragraph 32 urges the establishment of a management body for GORA to deliver the services in affordable housing.

4.2.2 State-Owned Company (SOC) as GORA Management

In line with the global privatization of social housing, the Government of Indonesia establishes PERUMNAS (National Housing Company-NHC) as a State-Owned Company (SOC) in the mid-1970s. It bears a mandatory obligation to provide affordable housing for meeting the housing needs nationwide.

However, this obligation poses contradictory tasks. Company Law 4/1962, as the legal basis for SOC, designates it as one of the spearheads to increase economic production, job opportunities, and government revenue. After the enactment of Municipal Law 23/2014, SOC plays a pivotal role to increase the municipal budget revenues to boost development. Government Regulation 54/2017 also solidifies the role of SOC is one of the spearheads to boost economic development and increase the state revenue. Since then, its role as a profit-oriented entity becomes the main priority and disregards its role of welfare-oriented [15]. Therefore, utilizing SOC to provide affordable housing will result in antinomy, as it is a profit-oriented, not a welfare-oriented agency [16].

This contradiction is demonstrated in several prior studies in Indonesia. A study in Cengkareng GORA (West Jakarta) shows the obligation poses a dilemmatic task for NHC, as it simultaneously must provide affordable housing and revenue. This study also describes the demand from low-income tenants to the municipal government for subsidizing their affordable rental fee, as a form of compensation after relocation from their previous high-density kampung settlement [17]. A thorough study in Indonesia asserts the importance of GORA to meet the housing need of low-income households, who have been excluded from the capitalistic housing market system. This study shares compelling evidence of how affordable housing provision by the private sectors fail to accommodate the housing needs for the designated beneficiaries, but the higher income group [18].

4.2.3 Public Service Agency (PSA) as GORA Management

GORA management, as a work unit under the Public Housing and Settlement Area Office of DKI Jakarta Province, is in charge of managing the GORA in DKI Jakarta. It is a work unit that provides public services to the low-incomes. In managing GORA, it takes operational costs for the management, one of which is for the maintenance. The operational costs of managing the flat are derived from the rental rate retribution income. The problem of high flat rental arrears affects the operational management of GORA so that the Government of DKI Jakarta annually issues huge subsidies for the maintenance of the GORA since relying on the income from the rental rates is not sufficient.

A study in Mukakuning (Riau Island) addresses the rigorous state budgeting system that plays a pivotal role in this problem, as the maintenance and operational costs are calculated and proposed a-year earlier without any flexibility for the miscellaneous activities. This study suggests the PSA as a new managerial form for GORA, instead of SOC or municipal government, instead of the profit-oriented SOC [19].

The change in the financial management pattern of GORA management to PSA can be a way out to overcome the burden of the municipal budget expenditure in providing operational subsidies for the annual GORA management. This is because PSA has the flexibility to manage its resources to improve the services provided to the public. By implementing PSA, GORA management can carry out good business practices based on principles of economic and productivity to improve services for GORA residents.

This type of institutional form meets the suggestion by an authors for self-sufficient management but still prioritizes a welfare-oriented [5]. As the self-sufficient management, PSA for GORA management may evolve to become the housing authority [8], which promises a successful result to preserve social housing for the low-incomes in the capital city.

5 Conclusion

Public Service Agency (PSA) is considered a suitable institutional form for GORA management. This form allows GORA management to generate income for meeting the maintenance and operational costs, also potentially subsidize the occurred arrears and improve the service.

Nonetheless, this form is not implemented yet, which its success is not yet proven. Therefore, a pilot study should be delivered to examine the potential success and failure for overcoming the unresolved issue. More studies in PSA development and strategic financial planning are required to prepare for its implementation in the future. Furthermore, related regulations should be studied and formulated to ensure the established operation of PSA as GORA management in the future.

Acknowledgements We would like to express our sincere gratitude to the Directorate of Research and the Community Engagement University of Indonesia for funding this study to the conference under contract: NKB-3876/UN2.RST/HKP.05.00/2020.

References

1. Central Bureau Statistics (Badan Pusat Statistik Provinsi) DKI Jakarta (2019) Statistik Daerah Provinsi DKI Jakarta 2019. BPS, Jakarta

2. Public Housing and Settlement Agency (Dinas Perumahan Rakyat dan Kawasan Permukiman) DKI Jakarta (2019) Buku Data dan Analisis RP3KP. DPRKP, Jakarta
3. Central Bureau Statistics (Badan Pusat Statistik Provinsi) DKI Jakarta (2017) Pengeluaran Rata-rata Per Kapita perbulan Menurut Kelompok Bukan Makanan di Provinsi DKI Jakarta (rupiah). Retrieved from Badan Pusat Statistik Provinsi DKI Jakarta: https://jakarta.bps.go.id/statictable/2017/03/13/198/pengeluaran-rata-rata-perkapita-perbulan-menurut-kelompok-bukan-makanan-di-provinsi-dki-jakarta-rupiah2015.html
4. Pusat Data dan Informasi (2020) Rekapitulasi Total Tunggakan Retribusi Rusunawa Provinsi DKI Jakarta. Unpublished Report. Dinas Perumahan Rakyat dan Kawasan Permukiman DKI Jakarta
5. Oyebanji AO (2014) Development of a framework for sustainable social housing provision (SSHP) in England. University of Central Lancashire, Preston
6. Mullins D, Czischke D, van Bortel G (2012) Exploring the meaning of hybridity and social enterprise in housing organisations. Hous Stud 27(4):405–17
7. Czischke D, van Bortel G (2018) An exploration of concepts and polices on 'affordable housing' in England, Italy, Poland and The Netherlands. J Hous Built Environ https://doi.org/10.1007/s10901-018-9598-1
8. Lin J (2011) The development of affordable housing: a case study in Guangzhou city, China. Master thesis, KTH, Stockholm
9. Snyder H (2019) Literature review as a research methodology: an overview and guidelines. J Bus Res 104:333–339
10. Davis J, Mengersen K, Bennett S et al (2014) Viewing systematic reviews and meta-analysis in social research through different lenses. Springer Plus 3:511. https://doi.org/10.1186/2193-1801-3-511
11. Pusat Data dan Informasi (2020) Demografi Warga Rusunawa DPRKP 20 Juli 2020. Unpublished report. Dinas Perumahan Rakyat dan Kawasan Permukiman DKI Jakarta
12. Setiadi HA (2015) Analisis Faktor Berpengaruh terhadap Kepuasan Penghuni Rumah Susun Sewa: Studi Kasus Rumah Susun Sewa Kemayoran (Analysis of influential factor toward public housing tenant's satisfaction case study In Kemayoran Public Housing). Jurnal Permukiman 10(1):19–36
13. Ganisen S, Mohammed AH, Nesan LJ et al (2014) Critical success factors for low cost housing building maintenance organization. Jurnal Teknologi 74(1):31–40
14. Ramadhani AM, Yusuf MA, Kisnarini R et al (2017) The effectiveness of rental housing finance for low income households In Sombo Rental Flats, Surabaya. Int J Sci Technol Res 6(7):137–143
15. Jaya WK (2007) Policy Brief Badan Usaha Milik Daerah (BUMD). Dalam Studi Potensi Pendapatan Asli Daerah Provinsi Daerah Istimewa Yogyakarta. Pusat Studi Ekonomi dan Kebijakan Publik (PSE-KP) UGM, Yogyakarta
16. Wantu FM (2007) Antinomi dalam penegakan hukum oleh hakim. Mimbar Hukum 19(3):335–485
17. Subkhan M (2008) Pengelolaan Rumah Susun Sederhana Sewa di Cengkareng, Jakarta Barat. Unpublished Master Thesis. University of Diponegoro
18. Pusat Perencanaan Pembangunan Hukum Nasional (2013) Analisis dan Evaluasi Peraturan Perundang-undangan tentang Perumahan Rakyat. Kementerian Hukum dan Hak Asasi Manusia, Jakarta
19. Lumbantoruan RP, Ritohardoyo S (2014) Perspesi Penghuni terhadap Sistem Pengelolaan GORA Mukakuning di Kota Batam. Jurnal Bumi Indonesia 3(4)

Deep Learning Models for Intrusion Detection in Wi-Fi Networks: A Literature Survey

Achmad Eriza Aminanto and Muhamad Erza Aminanto

Abstract Recently, the number of devices that are connected to the Internet are increasing exponentially due to the rise of the Internet of Things (IoT) era. Despite many advancements of the IoT era, we have been exposed to cyber security threats. Moreover, in this Covid-19 pandemic situation, the trend of cyber crimes is also increasing sharply. In this paper, we discuss one of possible countermeasures to combat cyber threats, namely Intrusion Detection Systems (IDS). IDS usually leverage many different types of machine learning models to detect the unknown attacks. In order to avoid confusion for future researchers in this field, we examine several states of the art papers which leverage deep learning for IDS in Wi-Fi networks. For this purpose, we choose one common Wi-Fi networks dataset, called AWID dataset. By examining the recent studies, we are able to understand current problems of IDS in Wi-Fi networks and able to prepare the best machine learning model for the corresponding problem to achieve a safe environment with minimal risk of cyber threats.

Keywords Intrusion detection system · Deep learning · AWID dataset · Anomaly detection

1 Introduction

One of the biggest media in Japan, the Japan Times [1], has reported that four promising technologies these days are big data, blockchain, AI (artificial intelligence), and IoT (internet of things). Presented in [2] and [3], approximately 30 million hosts will be connected to the Internet, output about US$ 175 billion in 2020 all over the world. At the same time, based on [4], the South Korean Smart

A. E. Aminanto (✉) · M. E. Aminanto
School of Strategic and Global Studies, Universitas Indonesia, Depok, Indonesia
e-mail: eriza@sci.ui.ac.id

M. E. Aminanto
e-mail: erza.aminanto@ui.ac.id

L. Yola et al. (eds.), *Sustainable Architecture and Building Environment*,
Lecture Notes in Civil Engineering 161,
https://doi.org/10.1007/978-981-16-2329-5_14

Home Energy Management segment's revenue is estimated to US$ 600 Million by 2024. However, the challenges [5] are cyber-security concerns and privacy breach. In Indonesia, the trend of cyber crime continues to increase every year. This cyber crime is increasingly becoming a major concern for the police of the Republic of Indonesia, especially after the Covid-19 pandemic. More and more users surf in cyberspace, especially social media which is increasing significantly. However, technological sophistication is currently being misused as a medium to commit crimes, especially cyber crimes. Quoted from digital media antaranews [6], which compiles from the National Police Criminal Investigation Unit [7], the trend of cyber crime continues to increase and the majority of cybercrimes are dominated by fraud. National Police also held a content contest regarding cyber crime awareness [8]. This competition shows the seriousness of the Police in dealing with cyber crime and means that cyber crime is currently becoming a very important concern for the Police.

IoT environments will mostly consist of small, light battery devices, and interconnected via wireless medium. Unfortunately, they will not execute computationally-high tasks such as cryptography computation because it will drain the battery instantly [9]. The huge amount of sensitive data transmitted in the air is appealing for attackers, yielding the IoT environment a main target of cyber-crime. The rapid spread of IoT-enabled devices has caused wireless networks to expose passive and active attacks, and their number has increased dramatically [10]. The wide and rapid spread of computing devices using Wi-Fi networks has produced complex, large and high-dimensional data, which can cause confusion when capturing attack attributes and force us to strengthen system security measures. Intrusion Detection System (IDS) is one of the most common components in every network security infrastructure [11]. However, it is a difficult task to develop an IDS with autonomous machine learning functions in an IoT environment.

Machine-learning has been leveraged broadly in the IDS research topic. IDS use machine learning models as the classification algorithm because of their model-free properties and capability to be learned [12]. By using the state of the art of machine-learning models namely deep learning, we can achieve impactful benefits for improving current IDSs in Wi-Fi networks [13].

We recognize that since a lot of previous studies used different machine learning models, there is confusion about how to properly adopt deep learning in IDS applications, particularly in Wi-Fi network attacks. Some studies only use deep learning methods in a partial sense, while other studies use shallow networks. The complexity of deep learning methods may be one of the reasons. In addition, deep learning methods require a lot of time to train properly. However, we found that some researchers have adopted deep learning methods throughout IDS. We compared the performance of several top-performance IDS in Wi-Fi networks. By comparing these IDSs, we expect other researchers who want to do research in this area can understand current situation and problems in the field of IDS in Wi-Fi networks.

2 IDS Background

Normally, we can separate IDS into misuse and anomaly detection [14]. The first model uses a rule to make use of actual portrayals to monitor the network. This model is also regularly referred to as a signature-based model. This model intends to recognize exact matching to the database. Despite the fact that this model is the most utilized everywhere [15], this model has evident drawbacks. The essential weak point is that it can't distinguish vague assaults on the grounds that it just thinks about the regarded traits of the assault. So as to preserve the exhibition of misuse detection, we have to refresh the assault signature without fail, which is problematic. Furthermore, attackers usually create an attack that does not consolidate previous assaults [16]. Such assaults make it extra tough to create fitting marks for the misuse detection. Then again, the focal point of inconsistency cognizance is to pick out irregular motion designs in the watched statistics [15]. Anomaly detection model for the most section control measurable examination and data mining techniques [17]. Since the arrangement model has the normal ability to do away with interruption examples and facts in the instruction stage, it can become aware of new types of assaults except in the past information.

Two normal strategies are typically utilized in IDS, to be particular clustering and classification undertakings. In the preliminary step, it is troublesome and costly to acquire infinite named community association information for regulated preparing. In modern years, clustering examinations have come to be a common anomaly discovery strategy [17]. Clustering itself is a solo information investigation technique that partitions a bunch of unlabeled records designs into numerous groups or corporations so the examples in the group are like one another, on the other hand not the same as the examples of exclusive clusters [17]. Simultaneously, classification is a method that acknowledges generous and vindictive offers based on the given information (commonly from clustering results). Clustering and classification can be done without difficulty utilizing AI strategies.

3 Dataset

For the purpose of this research, we choose one common dataset in Wi-Fi networks. This dataset released by Kolias et al., so-called Aegean Wi-Fi Intrusion Dataset (AWID) [18]. This dataset comprises two different types, the first type called "CLS" which has four labels, while another type called "ATK" which has 16 labels. Actually, the 4 labels in the CLS are big groups of 16 labels in the ATK. For example, simulated attacks in the CLS dataset consists of the Caffe-Latte, Hirte, Honeypot, and EvilTwin attack types listed in the ATK. The AWID dataset can also be divided into a full and partial dataset. In the partial one, there are 1,795,595 instances for training data, which comprises 1,633,190 normal instances and 162,385 attack instances. Meanwhile for testing data, they provided 575,643

Table 1 Class data distribution [18]

Type	Class	Training	Testing
Normal	Unbalanced	1.633.190	530.785
	Balanced	163.319	53.078
Attack	Impersonation	48.522	20.079
	Flooding	48.484	8.097
	Injection	65.379	16.682
	Total	162.385	44.858

instances in the partial dataset, consisting of 530,785 normal instances and 44,858 attack instances. This dataset contains 155 attributes such as frame length, radio tap present flag, radio tap data rate, wlan version, etc. Some data are presented in numerical and categorical form (Table 1).

4 State-of-the-Art IDS Using DL

In this section, we examine several state of the art publications which are the top performer in IDS for Wi-Fi networks. These publications can be regarded as the benchmark for next research in Wi-Fi networks attacks. At the end of this section, we summarize all research in the table.

Vaca et al. [19] proposed a WNIDS, which makes use of integrated learning. The advantages of integrated learning is that many basic learners are exploited to build a prediction model. By doing so, more accurate classification is expected. This paper also validated using AWID dataset. Wang et al. [20] examined various types of attacks in Wi-Fi networks using Deep Neural Networks (DNN) and Stacked Autoencoders (SAE). They achieved an impersonation detector in the AWID dataset.

Ran et al. [21] built a semi-supervised learning IDS model using ladder network. Their aim is to understand complex features with better discriminative ability for anomaly detection. The ladder network itself is constructed from Auto Encoders. Auto Encoder has the same number of neurons in both input and output. Ladder network used in this paper is composed of two sets of encoders, ordinary encoder and noise-encoder, and a decoder. They developed two steps of training, supervised and unsupervised steps [21]. AWID dataset is also used in this paper.

Parker et al. [22] proposed IDS for IoT infrastructures leveraging machine learning models. In this paper, they proposed two machine learning models, namely DEMISe and DETEReD. The first model combines feature extraction and mutual information. The Authors use two layers of stacked auto encoders concatenated with original attributes to be selected by mutual information method. This approach can be considered as filter based techniques since using mutual information for the feature selection step. THen the second machine learning model used is DETERed. The main difference is the wrapper based feature selection is used here.

Table 2 Publication comparisons of IDS that use AWID dataset

Paper	Dataset	Model	Classification type	Impersonation Acc (%)	Overall Acc (%)
Aminanto et al. [23]	AWID	FS+SAE	Binary	99.97	–
Parker et al. [22]	AWID	SAE+MI	Binary	98	–
Ran et al. [21]	AWID	DL-LN	Multi-class	89.32	98.54
Vaca et al. [19]	AWID	Ensemble	Multi-class	95.87	–
Wang et al. [20]	AWID	SAE+DNN	Multi-class	98.4	–

Aminanto et al. [23] developed an impersonation attack detector, D-FES using deep learning models. They used stacked auto encoder combined with other machine learning models. The stacked auto encoder is able to transform original attributes into different types of attributes that are more meaningful for the classification task. The extracted features are combined with original features to be selected by various feature selection techniques before classification. They successfully demonstrated the effectiveness of D-FES to detect impersonation attacks in AWID dataset.

From Table 2, we can see that Aminanto et al. [23] and Parker et al. [22] are focused to build the impersonation detector, while the rest three papers are multi class classification by distinguishing four labels provided in AWID. The best detection rate for impersonation class is Aminanto et al. [23] with 99.97%. While, the best overall detection rate is achieved by Ran et al. [21] with 98.54%. These numbers are benchmark data to be beaten for next research in IDS for Wi-Fi networks using AWID dataset.

5 Conclusions

We have examined several top performer IDS in Wi-Fi networks using state of the art machine learning models. As a summary, machine learning gives a significant impact in IDS research especially in Wi-Fi networks. Machine learning used as a feature engineering and classification step. In this paper we discuss many papers leveraging stacked auto encoder to extract more meaningful features from raw data. The output of this paper is the performance benchmark for next research in IDS for Wi-Fi networks, which are 99.97% in impersonation attack class and 98.54% of overall accuracy of four classes in AWID dataset. For future research, we believe that many ways will improve current work, such as decreasing the running time, making the computation decentralized to many devices, reducing false positive, and so on.

References

1. SHUSUKE MURAI. New Tokyo research center aims to boost Japan's fourth industrial revolution. Available on https://www.japantimes.co.jp/news/2018/07/05/business/tech/new-tokyo-research-center-aims-boost-japans-fourth-industrial-revolution/#.W2wwVigzbb0
2. Revenue of the internet of things in Japan from 2013 to 2020 (in billion U.S. dollars). Available on https://www.statista.com/statistics/512254/iot-revenue-japan/
3. Internet of Things (IoT) Connected devices installed base worldwide from 2015 to 2025 (in billions). Available on https://www.statista.com/statistics/471264/iot-number-of-connected-devices-worldwide/
4. iGateResearch. South Korea smart home market, volume, household penetration & key company analysis—forecast to 2024. Available on https://www.researchandmarkets.com/research/j22c7g/south_korea_smart?w=5
5. Laura W. South Korea smart home market 2017-2024: household penetration for smart home applications is expected to hit above 45%. Available on https://www.prnewswire.com/news-releases/south-korea-smart-home-market-2017-2024-household-penetration-for-smart-home-applications-is-expected-to-hit-above-45-300573827.html
6. Antara News Infographic. Cyber crime trend is on the rise. Available on https://www.antaranews.com/infografik/1571604/tren-kejahatan-siber-meningkat
7. Media Indonesia. Criminal investigation unit of POLRI: number of cyber crimes increasing since January. Available on https://mediaindonesia.com/read/detail/322327-bareskrim-polri-jumlah-kejahatan-siber-meningkat-sejak-januari
8. Welcoming Bhayangkara Anniversary. Bareskrim holds "Cyber Police Festival". Available on https://www.antaranews.com/berita/1565120/sambut-hut-bhayangkara-bareskrim-gelar-cyber-police-festival
9. Sforzin A et al (2016) RPiDS: Raspberry Pi IDS—a fruitful intrusion detection system for IoT. In: 2016 international IEEE conferences ubiquitous intelligence & computing, advanced and trusted computing, scalable computing and communications, cloud and big data computing, internet of people, and smart world congress (UIC/ATC/ScalCom/CBDCom/IoP/SmartWorld). IEEE
10. Kolias C, Stavrou A, Voas J, Bojanova I, Kuhn R (2016) Learning internet-of-things security "hands-on". IEEE Secur Priv 14(1):37–46
11. Kolias C, Kambourakis G, Maragoudakis M (2011) Swarm intelligence in intrusion detection: a survey. Comput Secur 30(8):625–642
12. Sommer R, Paxson V (2010) Outside the closed world: on using machine learning for network intrusion detection. In: Proceedings of symposium on security and privacy, Berkeley, California. IEEE, pp 305–316
13. Anthes G (2013) Deep learning comes of age. Commun ACM 56(6):13–15
14. Saxe J, Berlin K (2015) Deep neural network based malware detection using two dimensional binary program features. In: 2015 10th international conference on malicious and unwanted software (MALWARE). IEEE, pp 11–20
15. Laskov P, Düssel P, Schäfer C, Rieck K (2005) Learning intrusion detection: supervised or unsupervised? In: International conference on image analysis and processing. Springer, Berlin, Heidelberg, pp 50–57
16. Zanero S, Savaresi SM (2004) Unsupervised learning techniques for an intrusion detection system. In: Proceedings of the 2004 ACM symposium on applied computing, pp 412–419
17. Tsang CH, Kwong S (2006) Ant colony clustering and feature extraction for anomaly intrusion detection. In: Swarm intelligence in data mining. Springer, Berlin, Heidelberg, pp 101–123
18. Kolias C, Kambourakis G, Stavrou A, Gritzalis S (2015) Intrusion detection in 802.11 networks: empirical evaluation of threats and a public dataset. IEEE Commun Surv Tutorials 18(1):184–208

19. Vaca FD, Niyaz Q (2018) An ensemble learning based wi-fi network intrusion detection system (wnids). In: 2018 IEEE 17th international symposium on network computing and applications (NCA). IEEE, pp 1–5
20. Wang S, Li B, Yang M, Yan Z (2018) Intrusion detection for WiFi network: a deep learning approach. In: International wireless internet conference. Springer, Cham, pp 95–104
21. Ran J, Ji Y, Tang B (2019) A semi-supervised learning approach to IEEE 802.11 network anomaly detection. In: 2019 IEEE 89th vehicular technology conference (VTC2019-Spring). IEEE, pp 1–5
22. Parker LR, Yoo PD, Asyhari TA, Chermak L, Jhi Y, Taha K (2019) DEMISe: interpretable deep extraction and mutual information selection techniques for IoT intrusion detection. In: Proceedings of the 14th international conference on availability, reliability and security, pp 1–10
23. Aminanto ME, Choi R, Tanuwidjaja HC, Yoo PD, Kim K (2018) Deep abstraction and weighted feature selection for Wi-Fi impersonation detection. IEEE Trans Inf Forensics Secur 13(3):621–636

Sustainability Indicators for Biogas Production from Palm Oil Mill Effluent: A Case Study in Indonesia

Fentinur Evida Septriana, Tri Edhi Budhi Soesilo, and Ahyahudin Sodri

Abstract The sustainability assessment of the biogas production from palm oil mill effluent is complex, so it requires indicators that represent the sustainability concept of the system. This study aims to determine sustainability indicators of the biogas production system from palm oil mill effluent using the participatory analytic hierarchy process. This study results in ten sustainability indicators, in order of importance, consisting of the availability of raw material, profitability, greenhouse gases emissions, energy efficiency, water usage, production cost, investment, installed capacity, residents benefits, and job creation. These indicators help evaluate the sustainability of the biogas production system from palm oil mill effluent in the Bangka Belitung Islands and other regions in Indonesia.

Keywords Sustainability indicators · Renewable energy system · Biogas production · Palm oil mill effluent · Analytic hierarchy process

1 Introduction

Biogas from Palm Oil Mill Effluent (POME) has been recognized as a renewable energy source that helps meet energy needs together with the wastewater treatment to reduce water pollution and greenhouse gases (GHG) emissions. In Indonesia, biogas production from POME has been carried out in several places, one of which is in the Province of Bangka Belitung Islands. Over the years, the palm oil industry together with the refined tin industry have dominated the processing industry sector which accounts for nearly a fifth of the gross regional domestic product in the Bangka Belitung Islands, becoming the largest contribution to the economic

F. E. Septriana · T. E. B. Soesilo (✉) · A. Sodri
School of Environmental Science, Universitas Indonesia, Jalan Salemba Raya 4, Jakarta, Indonesia
e-mail: soesilo@indo.net.id

A. Sodri
e-mail: ahyahudin.sodri@ui.ac.id

L. Yola et al. (eds.), *Sustainable Architecture and Building Environment*, Lecture Notes in Civil Engineering 161,
https://doi.org/10.1007/978-981-16-2329-5_6

structure [1]. Several palm oil mills that produce large amounts of POME are combined with biogas plants that produce biogas from POME to generate electricity sold to the State Electricity Company. It supports local government efforts to increase the use of electrical energy from renewable energy sources. The production of biogas from POME is a form of renewable energy system. Therefore, to facilitate sustainable development planning that considers the needs of the current and future generations [2], especially in renewable energy development, it is necessary to know the sustainability of the biogas production from POME.

The concept of sustainability is very appropriate to be applied to complex problems because it integrates cross-sectoral aspects and involves multidisciplinary sciences. Sustainability assessment is complex because it requires interacting multidisciplinary aspects [3]. Solving the complexity of the sustainability of renewable energy systems requires indicators that can reflect the integrity of the system and the interaction of its subsystems [4]. Assessment of the sustainability of renewable energy systems must be able to explain system performance [5], which can be done by evaluating parameters that represent the integral concept of sustainability [4]. Thus, these sustainability indicators are truly valid and reliable to measure the sustainability of renewable energy systems. The main objective of sustainability indicators is to provide a framework for sustainability assessment that is driven by comprehensive, scalable, relevant, and easy to understand information [6].

Indicators are representative symbols used to explain trends in complex entities or systems [6], such as the sustainability of energy systems. Assessment of the sustainability of energy systems generally requires the use of resource, environmental, social, and efficiency indicators [4]. The use of various indicators is necessary to support the decision-making process related to energy sustainability policies. The sustainability of the renewable energy system, in principle, refers to the performance of the system in environmental, economic, and social aspects [7], so that at least these three aspects are involved in the sustainability assessment.

The process of selecting indicators requires parameters associated with reliability, compatibility, practicality, and measurement limitations [8]. Basically, the indicators for the sustainability of the renewable energy system must be able to reflect the concept of sustainability, measure the quality that is in line with sustainability objectives, be based on current and reliable information, reflect strategic views, provide a reference for system optimization, and reflect longevity of the system [4, 5].

Several studies on sustainability indicators of renewable energy systems have been carried out before. Studies in Indonesia have been carried out for bioenergy sustainability indicators [9], which was followed by recommendation of Indonesian Bioenergy Sustainability Indicators (IBSI) through several focus group discussions [10], and determination of sustainability indicators for palm oil-based bioenergy [11]. However, the specific sustainability indicators for the biogas production system from POME have not been determined. Furthermore, most of the studies on sustainability indicators of renewable energy were carried out in developed countries, even though the use of renewable energy was being increased in developing countries.

This study aims to determine the sustainability indicators of the biogas production from POME that can be used to evaluate the sustainability of the system through the model that will be developed. The results of this study are useful for assessing the sustainability of the biogas production system from POME in the Bangka Belitung Island, Indonesia.

2 Methods

Multi Criteria Decision Analysis (MCDA) is considered as the most suitable method for solving problems related to energy issues, considering that decision making on energy issues usually consists of many actors and many criteria that have several objectives [12]. Analytic Hierarchy Process (AHP) is one of the MCDA methods that breaks down general and uncontrolled problems into more specific and controlled sub-problems, then combines the solutions of all sub-problems into a conclusion [13]. Weighting the priority ranking using the AHP method is increasingly being used because of its easy-to-understand theory and simple application [8]. This study used AHP to determine the sustainability indicators for biogas production from POME.

This study was conducted in the Bangka Belitung Islands from August 2020 to October 2020. Sustainability indicators of renewable energy were identified from previous studies [5, 11, 14–16], then these indicators became the subject of focus group discussions (FGD). The indicators from the discussion results are then scored collectively by the FGD participants based on the criteria that have been determined in the discussion. Focus group discussions were held by involving representatives of stakeholders from two oil palm companies that have biogas plants, Energy and Mineral Resources Agency, Environment Agency, and State Electricity Company. The pairwise comparison matrices were used for weighting criteria and indicators. The weighting was carried out based on a rating scale of 1–9 according to Saaty [17]. In summary, the stages of the study taken are presented in Fig. 1.

3 Results and Discussion

The priority order of the sustainability indicators was determined based on the criteria determined in the FGD using a hierarchical structure as shown in Fig. 2. The criteria for the sustainability indicators to be assessed were determined through the FGD process, consisting of 7 criteria, namely relevance, measurability, informativeness, reliability, strategic, optimization, and long term. The sustainability indicators set for assessment through the FGD process consist of 10 indicators, namely investment, production cost, availability of raw materials, profitability, residents benefits, job creation, GHG emissions, energy efficiency, installed capacity, and water usage.

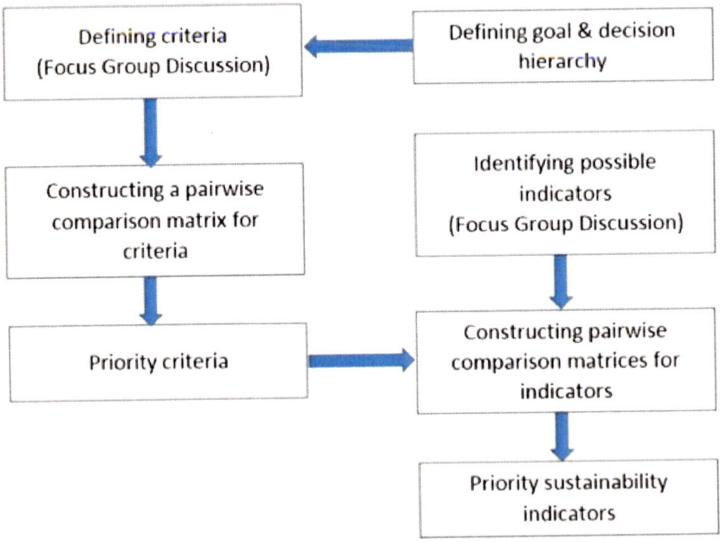

Fig. 1 The stages to determine sustainability indicators

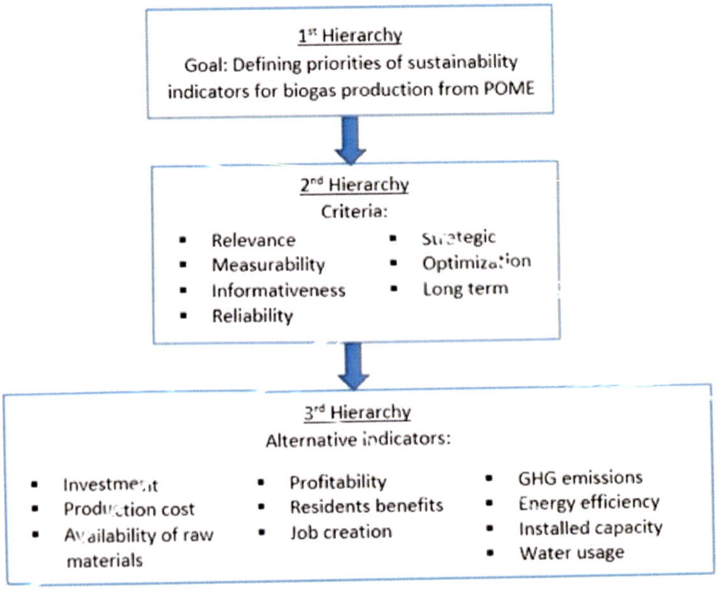

Fig. 2 Hierarchy structure to determine the priority sustainability indicators

The results of the criteria weighting with pairwise comparison matrix are presented in Fig. 3, with overall consistency ratio of 0.04. The "long term" criterion has the most important position compared to the other six criteria, followed by the criteria of relevance, measurability, reliability, informativeness, optimization, and strategic respectively. The "long term" criterion has the highest level of importance because the sustainability indicators of renewable energy must reflect the longevity of the system, in accordance with the concept of sustainability which shows the system's ability to survive, develop, and adapt to changes that occur in the long term [18].

Criteria with a lower level of importance were also needed to determine sustainability indicators at a later stage. Sustainability indicators must be relevant [5] to the renewable energy system being observed to be meaningful [19], easy to measure so that it facilitates the assessment process, reliable [5, 8], can inform the current system conditions (information is easy to obtain), can be used for system optimization [5], and strategic (related to several interests). The results of the criteria weighting in Fig. 3 were used to determine priority sustainability indicators which were also weighted using pairwise comparison matrices.

The results of indicators weighting for each criterion are briefly presented in Fig. 4. The consistency ratio of indicators weighting on the criteria of relevance, measurability, informativeness, reliability, strategic, optimization, and long term is 0.08, 0.04, 0.03, 0.05, 0.03, 0.05, and 0.03, respectively. The availability of raw materials has the highest level of importance compared to other indicators, in almost all criteria (except in "reliability" and "optimization"). In the "reliability" criterion, availability of raw materials has a slightly lower weight than production cost. This indicates that based on the stakeholders' view, production cost is more reliable in measuring sustainability than the availability of raw materials. In the "optimization" criterion, availability of raw materials has a slightly lower weight than energy efficiency. This indicates that based on the stakeholders' view, energy efficiency is more suitable to indicate optimization of the biogas production system from POME than the availability of raw materials. The results of overall weighting of sustainability indicators for the biogas production from POME, based on all criteria is shown in Fig. 5.

The sustainability indicator of renewable energy that was considered to be the most important by stakeholders is the availability of raw materials. This indicator is very important to measure the sustainability of the biogas production system from POME because based on their experience, the continuity of biogas production from

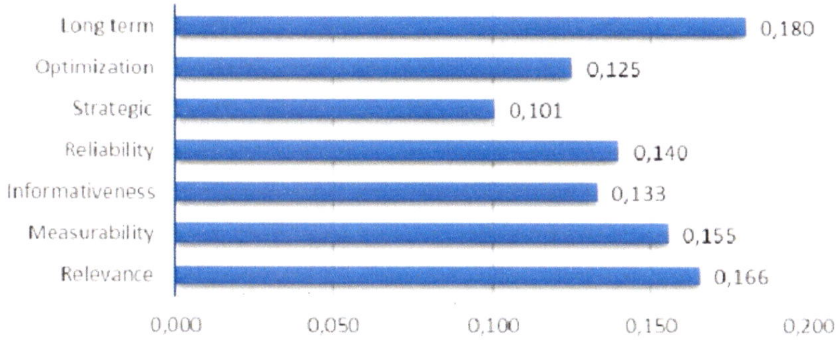

Fig. 3 Criteria importance weighting results

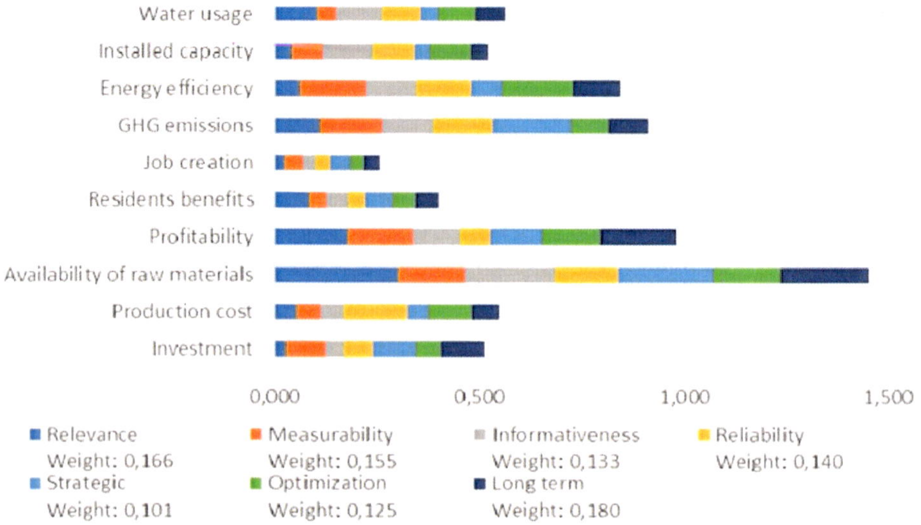

Fig. 4 The weights of the alternative indicators on each criterion

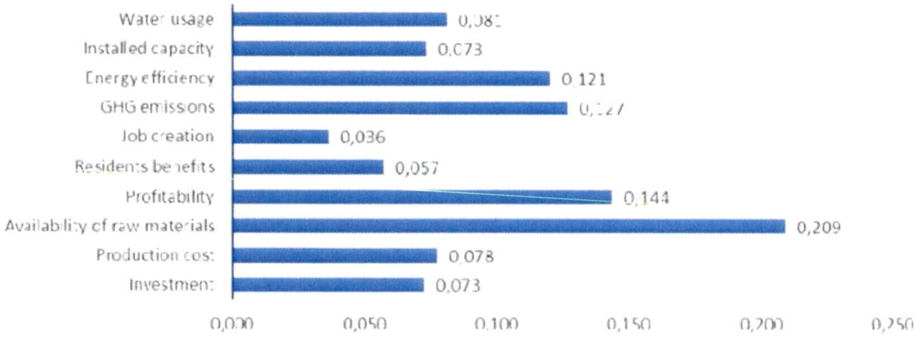

Fig. 5 Sustainability indicators importance

POME is very much influenced by the supply of POME from palm oil mills. The biogas production process cannot be sustainable if the available raw materials are very limited or only available at certain times. The production of biogas that is not continuous will cause the generation of electrical energy which is also not continuous and difficult to control. It will affect the sustainability performance of the biogas production system from POME as a whole. The availability of POME as raw material depends on the production process at the palm oil mill which process fresh fruit bunch. This raises the pros and cons of several parties because the fresh fruit bunch come from plantation land which is closely related to land use, regional spatial planning policies, issues of forest land conversion, biodiversity reduction, and greenhouse gas emissions.

The yield of oil palm plantations has decreased and has led to decreased production activities in palm oil mills in the Bangka Belitung Islands. As a result, POME, which is the raw material for biogas production, has also decreased. Thus, the palm oil industry also affects the availability of raw materials for biogas production from POME. Based on this condition, it can be understood if stakeholders view the availability of raw materials as the most important indicator of the sustainability of biogas production from POME.

Other indicators that relatively have high importance are profitability, GHG emissions, and energy efficiency. Profitability is one of the important factors affecting economic sustainability [16]. Profitability is important to be used as a sustainability indicator because it affects the ability to return capital. In addition, profitability can also reflect a company's economic ability to generate maximum profit, which is determined by the selling price and production costs. In the Bangka Belitung Islands, the selling price is largely determined by government policy factors and the sole buyer of electricity generated, namely the State Electricity Company, so that profitability is very important to be used as an indicator for measuring the sustainability of the biogas production system from POME, whose main function is to generate electricity.

In the environmental dimension, GHG emissions have generally been an indicator of the sustainability of renewable energy systems in various previous studies [11, 14, 15, 20–23] because it is closely related to the issue of world climate change. Biogas from POME is an anaerobic process product whose main component is methane gas so that biogas production is an effort to reduce GHG emissions in palm oil mills. The performance of the process in the plant is greatly influenced by energy efficiency. The efficient processes require lower capital and operating costs [14]. Energy efficiency can also reflect the wasting energy that occurs in the plant due to heat loss that is released into the environment. In a broad sense, renewable energy sustainability can be achieved if it is in line with the concept of energy efficiency [24], because the use of renewable energy and energy efficiency are two components that must be applied together [25].

The other six indicators (water usage, production cost, investment, installed capacity, residents benefits, and job creation) will still be used to assess the sustainability of the biogas production system from POME, although it has a lower weight. This weight will affect the portion of the system's sustainability assessment on the related indicators. The weighting results show the priority of indicators in the economic dimension over social dimension which indicate the stakeholders' priority in the formulation of policies planning related to the biogas production from POME. Based on routine reports on the implementation of environmental management and monitoring efforts by the company, the majority of residents around the site welcome the existence of the biogas plant (the plant location is quite far from residential areas), and many local residents are involved laboring in the biogas plant or palm oil mill. This reduces the concern of stakeholders to the social-community dimension. On the other hand, the selling price of electricity generated from biogas, which significantly affects the company's profitability, needs to be considered. As the results above, profitability significantly affects the sustainability of the biogas plant's process and the availability of raw materials that are highly dependent on operations in palm oil mills and plantations.

4 Conclusions

Sustainability indicators for the biogas production system from POME in the Bangka Belitung Islands were determined by the participatory AHP method, in order of importance, consisting of availability of raw materials, profitability, GHG emissions, energy efficiency, water usage, production cost, investment, installed capacity, residents benefits, and job creation. Based on the stakeholders' views, the availability of raw material is the most important indicator. The sustainability indicators will be used to assess the sustainability of the biogas production system from POME that will be modeled in further study, with the portion of the assessment being proportional to the weight of each indicator.

Stakeholders place the economic dimension above other dimensions, based on their experience which shows that economic factors often hamper the continuity of biogas production than other dimensions. This does not mean that other dimensions are not considered in the assessment of the sustainability of biogas production from POME, because the concept of sustainability is always based on at least three dimensions (environmental, social, economic) that influence each other in the long term.

Acknowledgements The authors would like to express gratitude for the financial support from the International Indexed Publication Grant (Hibah PUTI) Universitas Indonesia (contract number: NKB-2578/UN2.RST/HKP.05.00/2020) and the Provincial Government of Bangka Belitung Islands.

References

1. BPS (2020) Statistik Daerah Provinsi Kepulauan Bangka Belitung, Tahun 2020. BPS Provinsi Kepulauan Bangka Belitung, Pangkalpinang
2. World Commission on Environment and Development (1987) Our common future. Oxford
3. Sala S, Ciuffo B, Nijkamp P (2015) A systemic framework for sustainability assessment. Ecol Econ 119:314–325
4. Afgan NH, Carvalho MG, Hovanov NV (2000) Energy system assessment with sustainability indicators. Energy Policy 28(9):603–612
5. Liu G (2014) Development of a general sustainability indicator for renewable energy systems: a review. Renew Sustain Energy Rev 31:611–621
6. Hák T, Moldan B, Dahl AL (2007) Sustainability indicators: a scientific assessment. Island Press, Washington
7. Campos-Guzmán V, García-Cáscales MS, Espinosa et al N (2019) Life cycle analysis with multi-criteria decision making: a review of approaches for the sustainability evaluation of renewable energy technologies. Renew Sustain Energy Rev 104:343–366

8. Wang JJ, Jing YY, Zhang CF et al (2009) Review on multi-criteria decision analysis aid in sustainable energy decision-making. Renew Sustain Energy Rev 13(9):2263–2278
9. Food and Agriculture Organization of the United Nations (FAO) (2014) Pilot testing of GBEP sustainability indicators for bioenergy in Indonesia
10. Hambali E, Papilo P, Arkeman Y et al (2017) Indonesian sustainability bioenergy indicators, first edit. IPB Press, Bogor
11. Papilo P, Hambali E, Sitanggang IS (2018) Sustainability index assessment of palm oil-based bioenergy in Indonesia. J Clean Prod 196:808–820
12. Kumar A et al (2017) A review of multi criteria decision making (MCDM) towards sustainable renewable energy development. Renew Sustain Energy Rev 69(November 2016):596–609
13. Saaty TL (1990) How to make a decision: the analytic hierarchy process. Eur J Oper Res 48(1):9–26
14. Evans A, Strezov V, Evans TJ (2009) Assessment of sustainability indicators for renewable energy technologies. Renew Sustain Energy Rev 13(5):1082–1088
15. Ghenai C, Albawab M, Bettayeb M (2020) Sustainability indicators for renewable energy systems using multi-criteria decision-making model and extended SWARA/ARAS hybrid method. Renew Energy 146:580–597
16. Padilla-Rivera A, Paredes MG, Güereca LP (2019) A systematic review of the sustainability assessment of bioenergy: the case of gaseous biofuels. Biomass Bioenergy 125(October 2018):79–94
17. Saaty RW (1987) The analytic hierarchy process-what it is and how it is used. Math Model 9(3–5):161–176
18. Miller GT, Spoolman SE (2016) Environmental science, fifteenth. Cengage Learning, Boston
19. Dale VH, Kline KL, Richard TL et al (2018) Bridging biofuel sustainability indicators and ecosystem services through stakeholder engagement. Biomass Bioenerg 114:143–156
20. Cîrstea SD, Moldovan-Teselios C, Cîrstea A et al (2018) Evaluating renewable energy sustainability by composite index. Sustain 10(3)
21. García-Álvarez MT, Moreno B, Soares I (2016) Analyzing the sustainable energy development in the EU-15 by an aggregated synthetic index. Ecol Indic 60(2016):996–1007
22. Kemmler A, Spreng D (2007) Energy indicators for tracking sustainability in developing countries. Energy Policy 35(4):2466–2480
23. Nzila C, Dewulf J, Spanjers H et al (2012) Multi criteria sustainability assessment of biogas production in Kenya. Appl Energy 93:496–506
24. Oyedepo SO (2012) On energy for sustainable development in Nigeria. Renew Sustain Energy Rev 16(5):2583–2598
25. Riti JS, Shu Y (2016) Renewable energy, energy efficiency, and eco-friendly environment (R-E5) in Nigeria. Energy Sustain Soc 6(1):1–17

Systematic Review: Information for FM-Enabled BIM

W. N. Wan Siti Hajar, K. Syahrul Nizam, and Z. Nurshuhada

Abstract A comprehensive facility maintenances information as required is critical to ensure a smooth integration facilities management data with BIM system. During the transition from design and construction stage to facilities management stage, there has been a few issues concerning data loss and data integration. This study conducted analysis based on the existing literature systematically on information required by facilities management to support BIM execution plan for FM implementation. This review adopted guidelines by the PRISMA Statement (Preferred Reporting Items for Systematic reviews and Meta-Analyses) review method. As a result, 15 related studies were identified from two main databases namely Scopus and Web of Science. This study was divided into two sections. The first section reported on findings regarding LOD requirement to support geometrical information on the model. The second section discussed on non-geometrical information. There were three themes emerged from the first section and another four main themes and 15 sub-themes from the second section of this study. These themes were useful to facilitate the standardisation of FM information for the development of BIM model. Several recommendations were emphasised at the end of this study related to conduct more related and relevant qualitative studies, to review on data implementation framework to suggest on the best practice for the industry and to publish on a standard for systematic review method to guide on research synthesis in the context of facilities management.

Keywords Building information modelling (BIM) · Facilities management (FM) · Information exchange · Systematic review

W. N. Wan Siti Hajar
Faculty of Built Environment, University of Malaya, Jalan Universiti, 50603 Kuala Lumpur, Malaysia

K. Syahrul Nizam (✉)
Department of Building Surveying, Faculty of Built Environment, University of Malaya, Jalan Universiti, 50603 Kuala Lumpur, Malaysia

Z. Nurshuhada
Department of Quantity Surveying, Faculty of Built Environment, University of Malaya, Jalan Universiti, 50603 Kuala Lumpur, Malaysia

L. Yola et al. (eds.), *Sustainable Architecture and Building Environment*, Lecture Notes in Civil Engineering 161,
https://doi.org/10.1007/978-981-16-2329-5_10

1 Introduction

In recent years, there has been an exponential growth in ubiquitous digital tools to support building operational and maintenance activities within the full life cycle of the building at facilities management phase. Subsequently, it causes an increase in demand for accuracy and completeness facilities management data to operate the tools effectively and efficiently. Systematic data organisation will connect people, process, and technology in achieving the same objective of facilities management [1]. Moreover, adequate information and data storage for native data-intensive operations will lead to more efficient processes of operation and maintenance [2]. Hence, the potential of building information modelling (BIM) application for facilities management relies on the effectiveness of information management at the early stage. However, the current practice of BIM reuse for FM is obstructed by the discrepancy between the information demand with the supply at hand-over [3]. Besides, data transition from design and construction stage to facilities management stage is vulnerable to the potential risk of data loss. As a result, the final product of BIM model does not include all needed facilities information by personnel [4]. Eventually, the model invariably generates data without relevancy and purpose for the organisation [5], hence lead to ineffectiveness of BIM application for FM, especially in supporting daily basis operation with-in the broad spectrum of FM practice [6].

The study attempted to analyse comprehensively on the existing literature to identify crucial FM information requirements among key players in the construction industry for BIM application. This study adopted asset information requirement structure from standard ISO 19650-3 as a guideline in developing the datasets that can support FM services at all three level competencies area and eventually achieving the organizational goals. This section discussed the central question related to this review, while the second section explained in-depth on the methodology adopted for this systematic review. Next, the third section reported on the findings. The findings for this study consisted of two components. The first component recorded on the first research question regarding LOD requirement to support geometrical information on BIM, and the second part presented records analysis on non-geometrical information needs. The last section appraised on the findings to identify practical approach in capturing the information and imply further suggestion for research. On the whole, this study focused on variable group members related to BIMFM industry and aimed to assist stakeholders in the construction project (i.e. architects, engineers, contractors, owners, client, and facility managers) with the process of implementing and developing a practical BIMFM framework.

1.1 Facilities Management Roles and Objectives

Facilities management has been positioned as a profession with role to drive efficiencies and productivity in workplace with ultimate objective to support core business strategy. Facilities management is a profession with integration of multi-discipline approaches [7], to achieve the core business goals by maintaining, developing and adopting the buildings after design and construction [3] by acting as an "enabler" to the strategic planning of the organisation [8]. The management works on facilities operates with the objectivity to minimise building maintenance cycle costs while deliver the expected quality of the facilities simultaneously [9]. FM represents the holistic support system that connects services with its customers with the following flows, from information to material and services to finance [10]. Facilities Management roles are implemented at three hierarchical levels with operational at the base follow by tactical, and strategic subsequently [11]. Operational level is the fundamental competency area for any FM-related organisations. The four key elements of FM objective are operating, maintaining, improving, and adapting building and infrastructure assets while delivering a holistic support system to the goals of the organisation, inhabitant, owners, and facility personnel [12].

The scope of work for facilities management team can expand beyond the specified roles. The role and function of the profession is dynamic in nature [13]. The conventional roles incorporated three key subjects which were organizational culture, people and facilities/assets [8], meanwhile at present the emerging technology advancement has seen the FM roles maximise the productivity by diffusing technology uses into its activities while maintaining the cycle interface between these three elements. Ehrenberg [14] emphasised that the role of facilities management team within the project life-cycle expanded into creating organisational and strategic value which differs to the traditional facilities that mainly involved on technical maintenance. It is one way to keep the profession relevant over the time despite the shift to digital FM services. This leads to new perspectives of facilities management, from connecting people with the place to regulating the cycle between people, place, and technology.

However, there exists challenges in digitalization facilities management such as the application BIM as central information to support building operation and maintenance activities. Among the key challenges in implementing BIM for FM were lack of standard on complete dataset of information and uncertainty on the whole process to identify the required information [15]. Thus, this study attempted to examine, identify, and formalize on information requirement as recorded by experts in the industry in majority of relevant papers. At the end of this study, this effort to standardise information requirement aim to facilitate the owner on lists of information that they can refer to in stipulating information needs to support FM objectives that has been set forth at the early stage.

1.2 Format and Structure of Information

In traditional practice, varied formatted data provided as handover deliverables including in hard and soft copy format resulted into inefficient workflows at a later [16]. At present, the rapid pace of technological implementation for FM industry has seen an increase demand for interoperable data format between the digital tools. The FM information collected and recorded within BIM-FM database can serve as general knowledge that can benefit larger group of key players such as designers for reference on future projects [17].

According to standard ISO 19650-2, exchange information requirements can be derived from organisational, asset and project information requirements. The final requirements merged into a group of information consists of three varieties namely geometrical information, non-geometrical information, and documentation. In addition, the structure of information consists of two types namely structure and unstructured. While structured information could be recorded in various formats including spreadsheet-based and model-based formats, the unstructured information needs to be linked to graphical elements in BIM models using hyperlinks [2], with the exchange model is the current preferred output format. The model-based format represents the physical product of building information modelling consisted the information requirements of the owner [15]. The American Institute of Architects (AIA) Contracts Documents Committee emphasised on six Level of Development of a model which were LOD 100, LOD 200, LOD 300, LOD 350, LOD 400 and LOD 500. The components of LOD can be divided into two namely Level of Detail and Level of Information that reflects the level of geometrical information required and non-geometrical information required on the exchange model, respectively. The specification for LOD at the early stage will define the final product of BIM model for the use of facilities management. In practice, the contract and standard requirement recommended LOD 500 as Level of Development for facilities management that concentrates on field-verified and accurate representation model integrated with graphical and non-graphical information of size, shape, location, quantity, and orientation. LOD 500 utilization for construction industry in Malaysia was still at low and preceded by the private sector while public sector had only practised on LOD 300 [18].

The variation of information incorporated by a building information model can be categorised as an n-D model. Whereas LOD refers to the detail required to be included within 2D or 3D model, BIM dimension refers to process of linking specific type of data to an information model. The additional dimensions of data were added at each stage of building life-cycle to provide an enhanced overview to the owner on delivery process, costing, and maintenance process [19]. The dimensional aspects of 3D and above are simply efforts to integrate the model with more information and knowledge in enhancing the integration and coordination in the architecture, engineering, and construction (AEC) sector [17]. In general practice, the seventh dimensional model is adequate to address information required by facility management. The 7D aspect integrates asset data such as component status, specifications, operation and maintenance manuals, and warranty information into as-built model [20]. These list of asset data requirements can be grouped together and expanded further following standard ISO 19650. According to the standard, the aspects for asset information requirement (AIR) included managerial, commercial, and technical aspects. AIR is used to specify Asset Information Model (AIM) which aimed to support the strategic and day-to-day asset management processes (Fig. 1).

The bilateral data interoperability between BIM and FM system is realized by accurate input and output data format. BIM interoperability, according to [21], can be supported by input file format such as IFC and data structure specifications such as COBie. They are designated as an efficient workflow management in AEC industry between software [22]. In terms of IFC, it specifically codifies on geometric and data of the object [23], while COBie emphases on delivering asset data for non-geometric information such as data of equipment and spaces [24, 25]. Subsequently, IFC schema will be populated with data collected at different phases such as design, construction, and commissioning before IFC data model can be used to configure and map to COBie dataset to support specific needs by the owner or facility [26].

While the following schemes collected data at design, construction and commissioning stage, post-construction information should also be embedded on BIM to carry out a continuous effective facilities management role throughout the whole life-cycle of the building. This information is highly required particularly for the existing buildings. Conventional information database in the form of computer software used by facility manager such as Computerised Maintenance Management System (CMMS) included limited variation of information, mainly involving maintenance operation and schedules [26]. CMMS provides an analysed dataset required for facilities management team to maintain the building. The software covered activities at least for (a) the creation and maintenance of equipment/asset records, equipment/asset bill of materials and equipment/asset and work order history inventory control, (b) the creation, scheduling, execution and completion of work order, (c) the development and scheduling of preventive maintenance

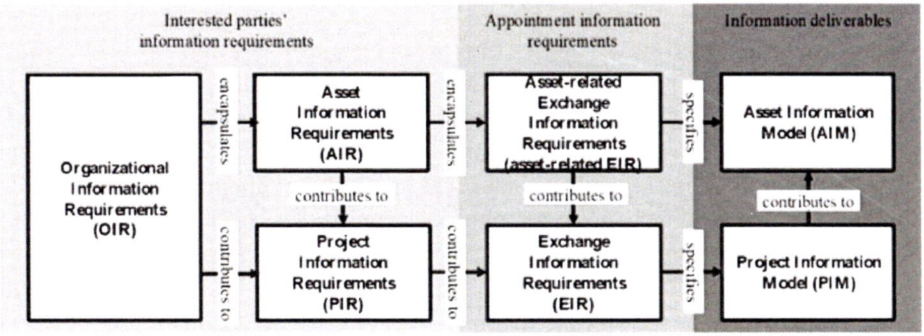

Fig. 1 Hierarchy of information requirements (adapted from Institution, 2020)

(PM) plan, (d) human resources, (e) the purchase and receive activity, (f) invoices matching and accounts payable and (g) table and reports [27]. The data can be mapped between the BIM models and CMMS database to allow for more effective scheduling of maintenance work order [28].

2 Methods

This systematic review adopted a rigorous searching strategy of scoping, planning, identification, screening, and eligibility [29]. The status of intended review had been searched and resulted into finding reviews with similar intention, however they were reviewed based on different research questions, served on different purposes, and adopted different approach. The following sub sections will explain in depth on the searching strategy to retrieve articles on the central question of information need and implementation for FM-enabled BIM framework. Based on the study above and research reviews, the categories of information requirement facilities management to integrate into building information modelling model set out into two, (a) LOD requirement for geometrical information and (b) Requirement for non-geometrical information.

2.1 PRISMA and Resources

This review referred and adopted guideline from the Preferred Reporting Items for Systematic Reviews and Meta-Analyses (PRISMA). PRISMA provides a rigorous searching strategy while maintaining records of relevant articles for reviewing. Three stages were involved systematically before the literatures were reviewed. At the first stage, a total of 785 records were identified by using the proposed query string on main databases (Table 1). 65 duplicated literatures were removed during the identification process. In addition to PRISMA guideline, the screening process were conducted at two stages due to masses record at the initial screening stage. At the first layer, the records were filtered out by refining articles results based on the inclusion criteria on the database. This resulted into exclusion of 498 records of conference proceedings book series, books, trade publications and non-English literature published before 2016. Furthermore, the second screening stage was reading the title and/or abstract of identified works from the searches. 179 articles were found did not meet the criteria while a total of 43 articles were included at the end of this process. Among the reasons for exclusion were articles did not focus on BIM for facilities management, required information for building operation activities were not included, and some were not empirical articles. Full-text version of 43 potential eligible articles were assessed in the next stage. At this stage, potential studies were sifted from the Method and Results for eligibility. The final record reported on 15 articles had passed the whole stage and eligible for the qualitative analysis (Fig. 2).

Table 1 Search strings used

Databases	Keywords used
Scopus	TITLE-ABS-KEY ((info* OR data* OR knowledge* OR deliver* OR require* AND specif* OR guidance* OR document* OR report*) AND (bim OR "build* info* model*" OR "*ated digital* model*" OR "*ated *D model*" OR "collaborat* *D Model" OR "collaborat* digital* model*") AND ("facilit* manage*" OR "facilit* maint*" OR "infrastructure* manage*" OR "infrastructure* maint*" OR "asset* manage*" OR "asset* maint*" OR "construct* maint*" OR "build* maint*" OR "operation* manage*"))
Web of science	TS = ((info* OR data* OR knowledge* OR deliver* OR require* specif* OR guidance* OR document* OR report*) AND (BIM OR "build* info* model*" OR "*ated digital* model*" OR "*ated *D model*" OR "collaborat* *D Model" OR "collaborat* digital* model*") AND ("facilit* manage*" OR "facilit* maint*" OR "infrastructure* manage*" OR "infrastructure* maint*" OR "asset* manage*" OR "asset* maint*" OR "construct* maint*" OR "build* maint*" OR "operation* manage*"))

MM mixed method, *QL* qualitative, *QN* quantitative

2.2 Eligibility and Criteria

The initial search strategy comprised of planning on search terms and search criteria. The relevant keywords were identified upon exploring related terms on encyclopedia, dictionary, thesaurus, and pertinent readings. The keyword identified were reflected on the search strings. Several criteria on inclusion and exclusion were developed to produce a definite search outcome. The criteria were based on source type, language, and timeline indexes. This review only considered research articles from journal and excluded literature reviews and any parts from book and conference proceedings. In addition, the records were limited to articles written in language English and Malay within a period of 6 years (2016–2020) as they served as adequate quantity and quality literatures for the study.

2.3 Data Abstraction and Analysis

This study aimed to conduct a rigorous analysis on preferred datasets to support building operational and maintenance activities using building information modelling from various literature to critically evaluate on each other characteristics. The review was focused on studies that responded to the questions formulated and their reliable methodology feature. At the initial stage, the data were assessed and analysed based on the abstract and conclusion before reading through the full articles (in-depth) to identify the related themes and sub-themes. Content analysis was set out for qualitative analyses in identifying themes related to FM information for building information modelling. The main themes for non-geometrical were formed with guidance from manual set out by BS EN ISO 19650-3 and sub-themes

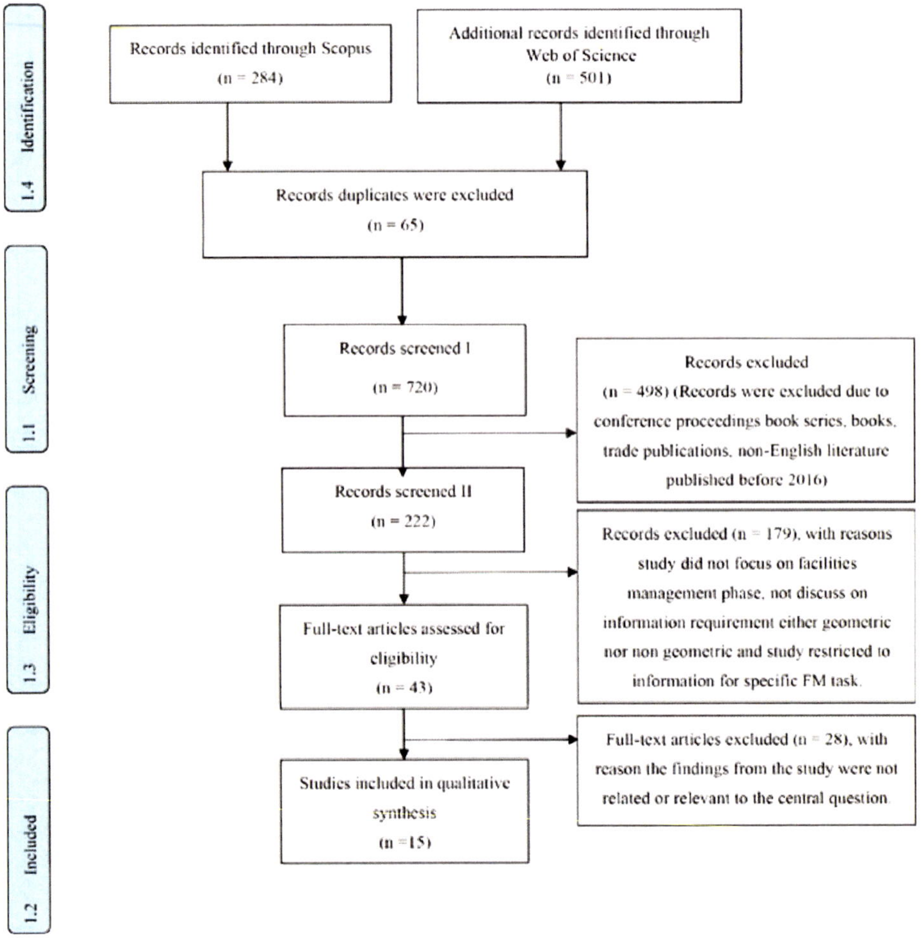

Fig. 2 Record appraisal process according to PRISMA guidelines

were developed from the themes identified from the literatures accordingly. The records adopted variety of methodology and multiple different findings development on the central question however, they were eligible for the analysis as they were related to fundamental truth which impacted in providing a greater and wider understanding on the problem statement [30].

3 Result

The central focus for this study was to identify information needs to enable BIM at FM phase. The result for this study was arranged into two sections. The first section presented geometrical data requirement analysed from a total of 8 studies from 15

Table 2 A total of 15 records identified for data abstraction

Studies	Country	Context of study	Main study design	LOD requirement	Non-geometrical requirement
[5]	US	Educational institutional	QL		√
[31]	Taiwan	General	QL	√	√
[17]	US	Facilities management activities in general	QN	√	√
[2]	US	Large educational institutional	QL	√	√
[15]		Educational institutional and public services infrastructure	MM	√	√
[42]	Vietnam	High-rise infrastructure in general	MM		√
[35]	China	General	MM	√	√
[43]	UK	Educational institutional for case study	QL		√
[39]	US	Educational institutional for pilot study	MM	√	√
[36]	Canada	Existing educational institutional and healthcare	QL	√	√
[44]	UK	General	MM		√
[32]	US	Educational institutional	QL	√	√
[33]	US	Existing educational institutional and public transit infrastructure	QL	√	√
[34]	Canada	General	QL	√	√
[45]		Utility tunnel	QL		√

records. The second section presented findings from 15 studies on non-geometrical data requirements, with four main themes and 15 sub-themes were identified (Table 2).

3.1 LOD Requirement

3.1.1 LOD Requirement Range Below LOD 500

Most of the current study supported on the use LOD lower than LOD 500 at FM stage. A smaller file size resulted into speedier modelling work [31]. To cut time consumed on remodelling building elements, this could be achieved over limited geometrical requirement such as LOD300 which included the exact location and dimensions [32]. Meanwhile, Abdirad and Dossick [33] opted for simplified visual cues of low-LOD interactive objects and shapes with facility data of asset location and systems connections of objects. In another study the highly preferred LOD requirement among FM personnel was LOD 350 to represent components in FM models and the highest LOD opted by the FM professional was LOD 400 [34].

3.1.2 LOD Developed from Lists of Detailed Requirement

Reference [31] defined LOD requirement from the detailed requirement of graphical and non-graphical for BIM model as assessed by FM experts. Among the important detailed information for geometrical requirement according to [35] were information for general building, building component, building site, product specification, types and location of building services, distribution and parameters of MEP and logical object tree organisation which matched with the basic model content requirements of LOD 500. Level of details for equipment maintenance requirements included specific requirements of assets, general guidelines from the owner or FM personnel and specific details of the project [17]. In another study, the new required geometric elements to support a specific end-user requirement were added to the base model of minimum geometrical information, regardless LOD requirements [36].

3.1.3 LOD Specified Based on Specific FM Function/Task

O&M personnel for maintenance, Basic Master Specification (BMS) system and asset management each required specific sets of information varied from each other at specific defined level of details for the same information [15, 34] identified specific LOD requirement to support specific FM task namely, corrective maintenance, preventive maintenance, asset management and space management. Meanwhile, certain assets parent-child relationship possessed their own asset data (such as air unit handling within fans) therefore they required specific level of development which can be distinct to each other [2].

3.2 Non-geometrical Requirement

This section focused on result for non-geometrical information needs. The analysis culminated in four main themes and a total of 15 sub-themes related to specific facility information for building operation. The four main themes are managerial, legal, technical, and commercial. Managerial information consists of seven sub-themes which are (i) identification of assets, (ii) location of the assets, (iii) spatial data related to assets, (iv) warranties and guarantee document, (v) historical maintenance record, (vi) schedule of maintenance and inspection tasks, and (vii) operational related plan, processes, and procedures. Legal information (two sub-themes) is (i) safety and disaster planning, and (ii) emergency planning. There are also two sub-themes in technical information namely (i) engineering specification and design parameters, and (ii) commissioning data. Meanwhile, commercial information consists of four sub-themes that are (i) specifications of assets, (ii) assets systems and classification, (iii) manufacturer and vendor data, and (iv) spare parts information.

3.2.1 Managerial Information

Managerial information had been considered by 15 studies as important facilities management information needs on BIM. The most crucial data for asset management are warranty and asset location (13 studies) in format model attribute data for operational activities. A total of 11 studies required operational related standards, processes, and procedures while 9 studies made use of identification of assets as requirement for their as-built model hand-over (Table 3).

3.2.2 Legal Information

In comparison to managerial information, legal information had been regarded as crucial by fewer studies, a total of 6 studies under two sub-themes. Safety and disaster planning featured in 4 studies, while there were 4 studies included emergency planning. Reference [5] considered both were equally essential to include in facility data needs.

3.2.3 Technical Information

Two sub-themes identified for technical information. A total of 8 studies reported facility data related to technical information as requirement for facilities management. Engineering specifications and design parameters were studied in 6 literatures while 5 records included commissioning data requirement.

3.2.4 Commercial Information

A total of 11 studies reported on commercial information in relation to information requirement for FM-enabled BIM. Under this themes, four sub-themes emerged, namely, asset description, the asset systems that they served, data related to manufacturer and vendor and spare part information such as identities and levels of spares held, interchangeability, specifications and storage locations.

Table 3 Findings on non-geometrical requirement

Studies	Managerial information							Legal information	
	Identification of assets	Location of assets	Spatial data related to assets	Warranties and guarantee document	Historical maintenance record	Future Schedule of maintenance and inspection tasks	Asset related standards, processes, and procedures	Safety and disaster planning	Emergency planning
[5]	√	√	√	√		√	√	√	√
[31]	√	√		√	√		√		√
[17]				√	√		√	√	
		√		√			√		
[15]	√	√		√	√	√	√		
[42]				√			√		
[35]	√	√	√	√	√	√	√	√	√
[43]	√	√	√	√	√		√		
[39]	√	√		√					
[36]		√	√		√	√	√		
[44]	√	√	√	√		√	√		
[32]		√		√			√		
[33]	√	√	√	√					√
[34]	√	√	√	√					
[45]		√						√	
Total	9	13	7	13	6	5	11	4	4

(continued)

Table 3 (continued)

Studies	Technical information		Commercial information			
Studies	Technical information		Commercial information			
	Engineering specification and design parameters	Commissioning data	Asset description, type and group	Assets systems	Manufacturer and vendor data	Spare parts information
[5]	√	√			√	√
[31]	√	√		√	√	
[17]	√					√
[15]		√		√	√	√
[42]						
[35]	√	√			√	
[43]			√		√	
[39]			√		√	
[36]						
[44]			√	√	√	√
[32]	√					
[33]	√		√	√	√	√
[34]		√			√	
[45]					√	
Total	6	5	4	4	10	5

4 Discussion

This study attempted to conduct a systematic review on existing records to analyse on information requirement for FM-enabled BIM. The identification of information needs to support facilities management on BIM has become an issue in the FM industry that resulted into a significant amount of resource waste therefore required a base guideline on requirement list for best practice. A rigorous record from two databases, Scopus, and Web of Science, resulted into 15 related articles that discuss the practice of BIM for FM industry across the world. The findings from the records indicated that variation of information is essential to achieve different requirement for each different organizational core strategy. At the first layer, the study intended to analyse on the LOD requirement to support facilities management on building information modelling. Further analysis of the study, four themes with 15 sub-themes emerged regarding non-geometrical information required to support building operational activities on BIM. The main themes were managerial to assist on decision-making, coordination, control, analysis and visualisation of organisational requirement, legal information for the law and legal process, technical information that concerns on all aspects of technicalities and commercial information related to the equipment, product, and other assets.

The findings for LOD requirement for facilities management highlighted on the need for a composite BIM Model rather than model that develop on one LOD fits all uses at facilities management stage. Aside from facilitating the owner in determine and identify the level of development of as-built model at hand-over, the impact of specifying only on required LOD for each FM task was that it leveraged the execution of BIM for FM with the supplication of manageable and operable files to use by FM personnel [37]. Based on the results of the analysed records, the highly preferred LOD for the industry uses range between LOD 300 to LOD 350. LOD 300 consists additional non graphical information which are scheduling and estimating data while LOD 350 includes additional information on details of elements interfacing with other buildings elements or systems. A query has arisen with majority experts' preference for lower LOD to facilitate FM tasks such as if LOD 500 for FM is yet to be undertook by the personnel within the industry or if there is lack of practicality for LOD 500 in the real context of the industry. This poses an area for future work to investigate on the practicality of LOD 500 for FM. The preferred application area of facilities management by using BIM is to locate the building components, accessing real time data, and for visual and marketing purpose [38]. This resulted a large quantity of empirical papers discussing on the asset location and crucial data to support operational and maintenance activities such as warranty, operational standards and procedures, asset ID and manufacturer data. Most of the following lists can be found on standard requirement such as COBie where the owner will specify on general COBie spreadsheet for close-out deliverables. COBie is a useful data format referred to data collected during design, construction, and commissioning phases however, there is a redundant amount of information in COBie for FM managers, hence, required a critical process of identifying information needed to deliver FM-enabled BIM [39].

The initial objective on the use of systematic review was to provide review with highly essential practical implications [40]. However, there were challenges in adapting this evidence-based methodology to the built environment to fully utilized the impact as identified in [41] such as fragmented and impracticality methodology, barrier on the comprehensibility of scientific evidence and difficulty in generalizing the findings that applied to particular setting considering the complexity division of built environment. With all effort, the author has conducted the following systematic review with rigorous searching technique and appropriate data abstraction and analysis as referred to PRISMA.

5 Conclusion

The following systematic review emphasised on the identification of crucial information to support facilities management on building information modelling. At present, research trends indicated an extensive research on BIM implementation for facilities management within the full life-cycle of the building. However, a dearth of research addressed the defining information requirements to support FM activities on BIM. Hence, the paper analysed previous literatures to construct themes information for FM-enabled BIM. The paper identified three main themes for the geometrical requirement at the first stage while the second stage of the study reported another four main classifications for non-geometrical information requirement. Themes for geometrical information were: (i) LOD requirement range below LOD 500, (ii) LOD developed from lists of detailed requirements and, (iii) LOD specified based on specific FM function/task. Meanwhile, at the second stage, the four main classification of information discussed were managerial, legal, technical, and commercial. These classifications were further specified into 15 sub-themes.

The study explored in detail the application of BIM at facilities management phase in variety sectors to identify aspects of the information requirement and its division that can be used to support building operational and maintenance activities. As there was lack of clear standardisation, the study facilitates the owners and facilities managers in developing the requirements according to the common practice within the industry and leverage the development of BIM model for FM uses. For this study, the themes developed on information requirements were limited to the two categories of information namely geometrical and non-geometrical while the further categories of information were not discussed for this study.

The review suggests several recommendations for future studies. First, more qualitative studies are needed by involving diverse profile of the organisations in the industry as it offers in-depth analysis and detailed information requirement based on specific organisational requirements. Second, to explore further on data implementation to suggest the most practical approach for the industry. Third, to have specific and a standard systematic review method to guide on research synthesis in context of facilities management information on building information modelling.

Acknowledgements This research is made possible with the support funding from Ministry of Education of Malaysia (KPM) through Fundamental Research Grant Scheme (FRGS)/1/2019/SSI11/UM/02/9.

References

1. Teicholz P, Mace B (2019) Integrating BIM and FM: status, case study and opportunities
2. Thabet W, Lucas J (2017) Asset data handover for a large educational institution: case-study approach. J Constr Eng Manage 143(11). https://doi.org/10.1061/(asce)co.1943-7862.0001389
3. Pärn EA, Edwards DJ, Sing MCP (2017) The building information modelling trajectory in facilities management: a review. Autom Constr 75:45–55. https://doi.org/10.1016/j.autcon.2016.12.003
4. Matarneh S, Danso-Amoako M, Al-Bizri S, Gaterell M, Matarneh R (2019b) BIM-based facilities information: streamlining the information exchange process. J Eng Des Technol
5. Mayo G, Issa RRA (2016) Nongeometric building information needs assessment for facilities management. J Manage Eng 32(3):04015054. https://doi.org/10.1061/(ASCE)ME.1943-5479.0000414
6. Teicholz PM (2013) BIM for facility managers [electronic resource]/IFMA, IFMA Foundation [Paul Teicholz]. Wiley, Hoboken
7. What is facilities management? Retrieved from https://www.ifma.org/about/what-is-facility-management
8. Kasim R (2011) Community-based facilities management theory and practice: Penerbit UTHM
9. Isa NM, Kamaruzzaman SN, Mohamed O, Jaapar A, Asbollah AZ (2016) Facilities management practices in malaysia: a literature review. MATEC Web Conf 66. https://doi.org/10.1051/matecconf/20166600054
10. Abdeen FN, Sandanayake YG (2018) Facilities management supply chain: functions, flows and relationships. Procedia Manuf 17:1104–1111. https://doi.org/10.1016/j.promfg.2018.10.074
11. Kamaruzzaman SN, Myeda NE, Zawawi EMA, Ramli RM (2018) Developing facilities management (FM) competencies for Malaysia. J Facil Manag
12. Pilanawithana Nethmin M, Sandanayake YG (2017) Positioning the facilities manager's role throughout the building lifecycle. J Facil Manag 15(4):376–392. https://doi.org/10.1108/JFM-06-2016-0024
13. Ikediashi DI, Ogunlana SO, Boateng P (2014) Determinants of outsourcing decision for facilities management (FM) services provision. Facilities
14. Ehrenberg M (2018) What is the future of facility management? FMJ 28(5):46–50
15. Cavka HB, Staub-French S, Poirier EA (2017) Developing owner information requirements for BIM-enabled project delivery and asset management. Autom Constr 83:169–183. https://doi.org/10.1016/j.autcon.2017.08.006
16. Tan AZT, Zaman A, Sutrisna M (2018) Enabling an effective knowledge and information flow between the phases of building construction and facilities management. Facilities 36(3/4):151–170. https://doi.org/10.1108/F-03-2016-0028

17. Liu R, Issa R (2016) Survey: common knowledge in BIM for facility maintenance. J Perform Constructed Facil 30(3). https://doi.org/10.1061/(asce)cf.1943-5509.0000778
18. Latiffi AA, Brahim J, Mohd S, Fathi MS (2015) Building information modeling (BIM): exploring level of development (LOD) in construction projects. Paper presented at the applied mechanics and materials
19. McPartland R (2017) BIM dimensions -3D, 4D, 5D, 6D BIM explained. Retrieved from https://www.thenbs.com/knowledge/bim-dimensions-3d-4d-5d-6d-bim-explained
20. Dallasega P, Marengo E, Nutt W, Rescic L, Matt DT, Rauch E (2015) Design of a framework for supporting the execution-management of small and medium sized projects in the AEC-industry
21. BIM Interoperability Standards. Retrieved from https://www.autodesk.co.uk/solutions/bim/bim-interoperability-standards
22. Eastman CM (2018) BIM handbook [electronic resource]: a guide to building information modeling for owners, designers, engineers, contractors, and facility managers [Charles E et al], 3rd edn. Wiley, Hoboken
23. Kensek K (2015) BIM guidelines inform facilities management databases: a case study over time. Buildings 5:899–916. https://doi.org/10.3390/buildings5030899
24. Guillen AJ, Crespo A, Gómez J, González-Prida V, Kobbacy K, Shariff S (2016) Building information modeling as assest management tool. IFAC-PapersOnLine 49(28):191–196. https://doi.org/10.1016/j.ifacol.2016.11.033
25. Hamil DS (2018) What is COBie. Retrieved from https://www.thenbs.com/knowledge/what-is-cobie
26. Florez L, Afsari K (2018) Integrating facility management information into building information modelling using COBie: current status and future directions. Paper presented at the 35th international symposium on automation and robotics in construction (ISARC 2018), Berlin, Germany
27. Aziz ND, Nawawi AH, Ariff NRM (2016) ICT evolution in facilities management (FM): building information modelling (BIM) as the latest technology. Procedia Soc Behav Sci 234:363–371. https://doi.org/10.1016/j.sbspro.2016.10.253
28. Chen W, Chen K, Cheng JCP, Wang Q, Gan VJL (2018) BIM-based framework for automatic scheduling of facility maintenance work orders. Autom Constr 91:15–30. https://doi.org/10.1016/j.autcon.2018.03.007
29. Siddaway AP, Wood AM, Hedges LV (2019) How to do a systematic review: a best practice guide for conducting and reporting narrative reviews. Meta-Analyses Meta-Synth 70(1):747–770. https://doi.org/10.1146/annurev-psych-010418-102803
30. Mays N, Pope C, Popay J (2005) Systematically reviewing qualitative and quantitative evidence to inform management and policy-making in the health field. J Health Serv Res Policy 10(1):6–20. https://doi.org/10.1258/1355819054308576
31. Lin Y-C, Chen Y-P, Huang W-T, Hong C-C (2016) Development of BIM execution plan for BIM model management during the pre-operation phase: a case study. Buildings 6(1):8
32. Sadeghi M, Elliott JW, Porro N, Strong K (2019) Developing building information models (BIM) for building hand-over, operation and maintenance. J Facil Manag
33. Abdirad H, Dossick CS (2020) Rebaselining asset data for existing facilities and infrastructure. J Comput Civ Eng 34(1):05019004
34. Dias PDR, Ergan S (2020) Owner requirements in as-built BIM deliverables and a system architecture for FM-specific BIM representation. Can J Civ Eng 47(2):215–227
35. Lu Q, Chen L, Lee S, Zhao X (2018) Activity theory-based analysis of BIM implementation in building O&M and first response. Autom Constr 85:317–332. https://doi.org/10.1016/j.autcon.2017.10.017
36. Bortoluzzi B, Efremov I, Medina C, Sobieraj D, McArthur JJ (2019) Automating the creation of building information models for existing buildings. Autom Const 105. https://doi.org/10.1016/j.autcon.2019.102838
37. Dias P, Ergan S (2016) The need for representing facility information with customized LOD for specific FM tasks. Constr Res Congr 2016:2563–2572

38. Becerik-Gerber B, Jazizadeh F, Li N, Calis G (2012) Application areas and data requirements for BIM-enabled facilities management. J Constr Eng Manag 138(3):431–442. https://doi.org/10.1061/(ASCE)CO.1943-7862.0000433
39. Pishdad-Bozorgi P, Gao X, Eastman C, Self AP (2018) Planning and developing facility management-enabled building information model (FM-enabled BIM). Autom Constr 87: 22–38. https://doi.org/10.1016/j.autcon.2017.12.004
40. Institution TBS (2020) BS EN ISO 19650-1: 2018—organization and digitization of information about buildings and civil engineering works, including building information modelling (BIM)—information management using building information modelling Part 1: concepts and principles
41. Hall S, Oldfield P, Mullins B, Pollard B, Criado-Pérez C (2017) Evidence based practice for the built environment: can systematic reviews close the—practice gap? Procedia Eng 180:912–924. https://doi.org/10.1016/j.proeng.2017.04.341
42. Nguyen PT, Nguyen TA, Ha NTH, Nguyen TN (2017) Facilities management in high rise buildings using building information modeling. Int J Adv Appl Sci 4(2):1–9
43. Farghaly K, Abanda FH, Vidalakis C, Wood G (2018) Taxonomy for BIM and asset management semantic interoperability. J Manage Eng 34(4). https://doi.org/10.1061/(asce)me.1943-5479.0000610
44. Matarneh S, Danso-Amoako M, Al-Bizri S, Gaterell M, Matarneh R (2019) BIM for FM: developing information requirements to support facilities management systems. Facilities 38 (5/6):378–394. https://doi.org/10.1108/F-07-2018-0084
45. Yin X, Liu H, Chen Y, Wang Y, Al-Hussein M (2020) A BIM-based framework for operation and maintenance of utility tunnels. Tunnel Underground Space Technol Incorporating Trenchless Technol Res 97. https://doi.org/10.1016/j.tust.2019.103252
46. Institution TBS (2013) PAS 1192-2 specification for information management for the capital/delivery phase of construction projects using building information modelling

Typology and Characteristics of Slum Settlement in Wetland of Palembang City

Hendi Warlika Sedo Putra, M. Baitullah Al Amin, and Ibrahim

Abstract Slum areas are one of the biggest problems in Indonesia. The city of Palembang is no exception, with a slum area reaching 2,472,981 ha making it one of the largest cities in Indonesia. It is the duty of the regional government to be able to complete it with a specific time target set by the central government. With limited time, there needs to be an effective strategy in solving the slums of Palembang city. The form of activity intervention based on the standard infrastructure of settlements is considered to have had a significant impact on the reduction of the slums of Palembang city. Typical urban characteristics that are located on the banks of rivers and swamps are their own problems that need to be studied further in order to formulate an activity program. Judging from the planning documents of the SIAP (Slum Improvement Action Plan) that have been prepared, the slums of Palembang city are divided into 2 regional typologies, namely slums in lowland and on the water's edge. It is considered inappropriate if we look at the definition and geographical conditions of Palembang city which are dominated by tidal areas of rivers and swamps. For this reason, this research was conducted to find out the types of typologies and characteristics of slums in Palembang City. Using quantitative methods based on supporting secondary data and basic data analysis, the overlay technique of slum maps and water areas is the basis for determining the typology of the slums of Palembang city. Then, a grouping of numerical data based on the typology of the processed area to be able to formulate characters from each typology of slum areas in the city of Palembang. From the results of the analysis, three typologies of slum areas were found in Palembang, namely the slums of on the water, on the water's edge, and lowlands.

Keywords Typology · Characteristic · Slum · Palembang · Settlement

H. W. S. Putra (✉) · Ibrahim
Department of Civil Engineering, Politeknik Negeri Sriwijaya, Palembang, Sumatera Selatan, Indonesia
e-mail: hendiwsp@polsri.ac.id

M. B. Al Amin
Department of Civil Engineering, Engineering Faculty, Universitas Sriwijaya, Palembang, Sumatera Selatan, Indonesia

L. Yola et al. (eds.), *Sustainable Architecture and Building Environment*, Lecture Notes in Civil Engineering 161,
https://doi.org/10.1007/978-981-16-2329-5_17

1 Introduction

The residential area is an element forming the urban environment. By definition, a settlement is a group of housing equipped with infrastructure, facilities, public and support utilities in both urban and rural areas. With the function of living and doing activities, everyone certainly hopes for a good residential environment. Fulfillment of housing infrastructure, facilities and public utilities as part of the settlement are an effort to fulfill a livable house [1].

The phenomenon in the urban settlement environment is the existence of slum settlements in several corners of the city. The slum settlement then becomes a problem faced in urban settlement areas. According to Law No. 1 of 2011 concerning Housing and Settlement Areas, slum settlements are described as uninhabited settlements caused by building irregularities, high levels of building density, and quality of buildings and facilities that do not meet the requirements [2]. The growth of slum environment in urban areas can be caused by direct factors and indirect factors [3]. Direct factors is more due to the physical condition of the slum environment. These factors include: factors of land area, factors of building types, and factors of facilities and infrastructure. Indirect factors emphasize that slum conditions are caused not only by environmental physical factors [1]. These factors include: factor of the number of residents, factor of occupancy status, factor of income, and factor of length of stay. Other than that, the high urbanization, illegal land placement in the urban environment, and the unfulfilled minimum service standard in urban areas are several factors that influence the growth of urban slum settlements [4]. The inadequacy of slum settlements is inseparable from the low quality of drinking water services, drainage, waste processing and sewage treatment systems, environmental accessibility as basic infrastructure in addition to building conditions that have an impact on the visual formation of slum areas.

Palembang as one of the big cities in Indonesia cannot be separated from the problem of slum settlements in urban areas. In accordance with what is stated in the Decree of the Mayor of Palembang number 488 in 2014 concerning the Determination of the Location of Slum Settlement Areas in Palembang, there are recorded as many as 2,472,981 ha of slum settlements spread across 59 sub-districts in Palembang. Palembang City as a river city has an area dominated by land which has an altitude of 3–4 m above sea level. However, rapid development made the construction of new settlements in wetlands inevitable [5]. Palembang with the characteristics of wetland or lowland areas, of course, has its own unique characteristics of problems and does not exist in all cities in Indonesia.

There are five basic elements of settlement: Nature, is land that can be used to build housing and other functions; Man, both individuals and groups who build or live in; Society, in which there are interactions and social relations between humans so as to form certain bonds as a society; Shells, namely buildings where humans live with their respective functions; Networks, in other words infrastructure that support environmental functions, both natural and man-made [4]. Other than that, the characteristics of a settlement can be captured by observers by conducting a study of: occupant characteristics, residential characteristics, facilities and infrastructure characteristics, environmental characteristics, and slum levels [4]. In the Regulation of the Minister of Public Works and Public Housing number 2 in 2016 concerning Quality Improvement of Slum Housing and Slum Settlements, in terms of several aspects elaborated through several indicators of assessment of slums [6]: Building, Neighborhood roads, Provision of drinking water, Environmental drainage, Waste water management, Solid waste management, and fire protection.

Reviewed from the planning documents that have been prepared by the city government of Palembang City, namely SIAP (Slum Improvement Action Plan) 2015–2019 document, the slum areas of the city of Palembang (59 sub-districts) fall into two typological categories of slum areas, namely slums in the lowlands and slums on the water's edge. The typology of slum housing and slum settlements is a grouping of slum housing and slum settlements based on geographic location [7]. The typology of slum housing and slum settlements consists of: Slum housing and slum settlements on water, Slum housing and slum settlements on the water's edge, Slum housing and slum settlements in the lowlands, Slum housing and slum settlements in the hills, Slum housing and slum settlements in disaster prone areas. When viewed from the environmental conditions (land topography/land character) of Palembang, which are generally river and tidal swamp areas, there is a big possibility that the typology of slum settlements in Palembang City is not in accordance with what has been formulated in the "SIAP" document. That is closely related to the typology definition of slum settlements on water located in the tidal area of swamps, rivers and sea [7].

Spatial information is important for several aspects of urban development, planning and management, including identifying and observing changes in slum settlements [8]. The result of research on "Typology and Characteristics of Slum Settlements in Wetlands of Palembang City" will produce an overview of the existing slum conditions in Palembang City, and can be used as a reference in carrying out actions and planning to prevent and improve the quality of the slum environment in Palembang City. In addition, it can also be used as a base map in further analysis related to slums of Palembang City.

2 Research Methods

The research method in general is more on quantitative research with a comparative study of data on slum settlements with the water conditions of Palembang City.

Data collection techniques are very dependent on the data and the type of data required: (a) Data on the distribution of slum areas in the city of Palembang, based on the Mayor's Decree number 488 of 2014, and the spatial data of KOTAKU Program in Palembang, (b) The numerical data on the characteristics of the slum characteristics of each sub-district in the city of Palembang is sourced from the updated data related to the KOTAKU program of the baseline data for slum settlements in Palembang, (c) Hydrological data or waters of Palembang City, which are obtained from secondary data from the Palembang City Development Planning Agency and also the results of analyzes related to the tide of Palembang City, (d) Base map of Palembang city and other support data.

Data processing; (a) Data on the distribution of slum in Palembang city, this data will be processed using computer programs such as AutoCAD/GIS/QGIS. Slum distribution data is in the form of mapping from secondary data of slum areas per neighbourhood from the results of the KOTAKU self-help mapping program related to slum settlements in 59 sub-districts in Palembang City, (b) Data on the area of water/topography of the city of Palembang. This data will also be processed in mapping related to the waters of the city of Palembang. The method used is mapping of inundation areas in lowlands based on an overlay between topographic data and forecasting of high tide in the Musi River. This analysis also considers the existing river network system, (c) Slum numerical data for settlement areas. In the KOTAKU program, there is basic data in the form of baseline data which will then be used as the basis for data processing to formulate the slum classification of each sub-district/settlement area, and also formulate the characteristics of slum in sub-district through a study of the slum aspect data and its parameters.

Data analysis is related to results of future research, namely the typology and characteristics of the slum areas of Palembang City; (a) The method used is to overlay the water data (tidal waters of Palembang City) with the data on the distribution of slum distribution in Palembang City using the QGIS program, (b) The slum characteristics of the residential area in Palembang City are carried out by grouping and processing numerical data that combines each area with a similar typology. Assessment of each parameter follows the rules (RP2KPKP guidelines), namely: 0–25% with a score of 0; 25–50% with a score of 1; 50–75% with a score of 2; and 75–100% with a score of 2. From the sum of the assessments of each parameter, it will produce a classification of slum level, namely the threshold <15 is declared not slum, the values 15–35 are mild slums, 36–55 are moderate slums, and 56–75 are heavy slums.

3 Results and Discussion

3.1 *Slum Area of Palembang City*

When viewed from the Mayor's Decree Number 488 in 2014, Palembang City has a total of 2,472,981 ha of slum area spread across 59 villages in 13 sub-districts. From the data collection carried out at the base level (per house/head of the family) through seven slum aspects as well as 19 slum parameters, the area of Palembang City Slum Area is 1,443,560 ha in 2017. From the intervention results of slum handling activities from several sources funding, in 2018 the area of slums of Palembang City became 1380.2 ha, which are spread across 59 sub-districts in Palembang City (Fig. 1).

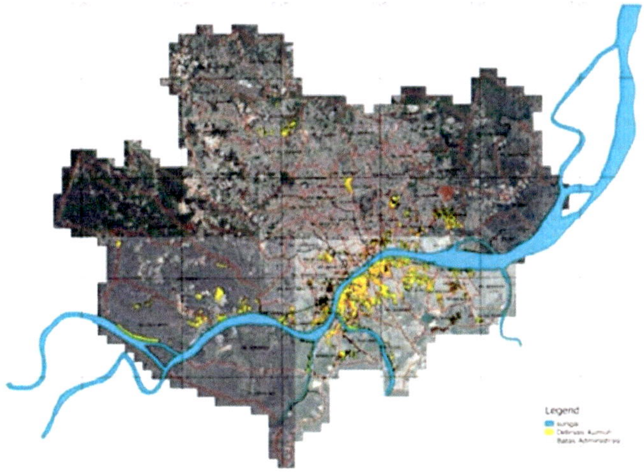

Fig. 1 Slum area of Palembang City

From a total of 59 sub-district, several classifications of slum level in the sub-district/area in Palembang City can be identified, namely: 1 sub-district with a Mild Slum classification with a slum area of 22.04 ha, 41 sub-districts classified as Medium Slum with a total slum area of 974.97 ha, and 17 sub-districts classified as Heavy Slum with a total slum area of 383.01 ha.

3.2 Water Territory of Palembang City

Palembang City is divided by Musi River into two large parts called Seberang Ulu and Seberang Ilir. Palembang City itself has as many as 108 tributaries, and there are 4 (four) large rivers in Palembang. The four rivers, namely Musi River, is the largest river with an average width of 504 m (the largest width is on the island of Kemaro which is 1350 m, and the smallest width is 250 m which is at the location of the Musi II Bridge). The other three major rivers are the Komering River with an average width of 236 m; Ogan River with an average width of 211 m; and the Keramasan River with an average width of 103 m. Apart from these large rivers, there are other small rivers located at Seberang Ilir which function as urban drainage (there are ±68 active tributaries). The small rivers have a width ranging from 3 to 20 m. In these rivers, retention ponds are built, so that they become part of the river boundaries.

The receding condition of the water area in Palembang city is made by overlaying the aerial image map of Palembang city with river network data (large rivers and small rivers) and swamp areas (conservation swamps, cultivation swamps, and reclamation swamps), which are calculated from the receding water conditions.

Tide conditions were carried out by modeling for 4 (four) major river basins in Palembang City, namely the Musi River, Komering River, Keramasan River, and Ogan River. Hydrograph analysis of flood water discharge to see inundation areas in the city of Palembang using 20 years return period. Tide forecasting is carried out using measured data in the Musi River (Sekanak estuary). From the results of this analysis, it was found that the average tide level in the city of Palembang was 3.3 m above sea level.

3.3 *Typology of Slum Areas in Palembang City*

Settlement typology analysis was carried out using the overlay method of Palembang city waters data with the existing slum area distribution data. Assessment of the position of the area to the water area is seen from the dominance of the neighborhood area in settlements within each typology (>50%), because from the results of the map overlay carried out, some data is obtained in the form of slices of each slum neighborhood, not all of them are within the scope of one slum typology of a settlement. From the analysis conducted, it was obtained an overview: (Fig. 2).

a. Slum typology based on low tide
   (1) Slum Settlements on Water, which are spread across 5 sub-districts and 7 neighborhoods in the area with a total slum area of 13,349 ha or 0.9% of the slum areas that are above Air in Palembang City. (2) Slum settlements on the waterfront, spread over 18 sub-districts with a total of 73 slum neighborhoods and an area of 111.125 ha of slum settlements or there are 7.7% of the slum

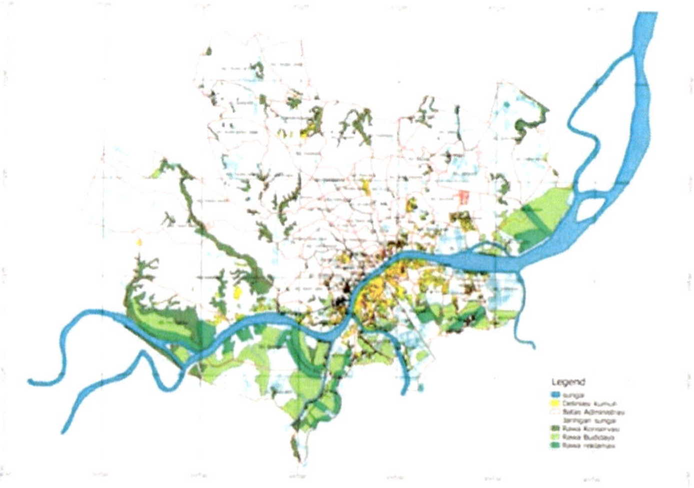

Fig. 2 Slum area overlay map—low tide level of Palembang

areas located on the Waterfront of Palembang. (3) Slum settlements in the lowlands, spread across 57 sub-districts of Palembang city and 785 neighborhoods with a total slum area of 1,319,082 ha or 91.4% of the slum areas found in the lowlands of Palembang City (Fig. 3).

b. Slum typology based on tide conditions
(1) Slum Settlements on Water, which are spread across 50 sub-districts and 517 neighborhoods in the area with a total slum area of 757,419 ha or 52.47% of the slum areas of Palembang City. (2) Slum settlements on the waterfront, spread over 6 sub-districts with a total of 9 slum neighborhoods and an area of 9023 ha of slum settlements or 0.63% of the slum areas of Palembang City. (3) Slum settlements in the lowlands, spread over 47 sub-districts of Palembang city and 339 neighborhoods with a total slum area of 677,114 ha or 46.91% of the slum areas of Palembang City.

Based on the typology definition of slum settlements on water located in the tidal area of swamps, rivers and seas, the tide conditions become a parameter in seeing the typology of slum areas in Palembang city. However, from the two regional analysis descriptions of low tide and high tide conditions, all of them show that there are three typologies of slum areas in the city of Palembang: on the water, on the water's edge, and in the lowlands.

3.4 Slum Characteristics

Physically, the character of the slum areas of Palembang City is characterized by unhealthy building and environmental conditions, a high level of building density,

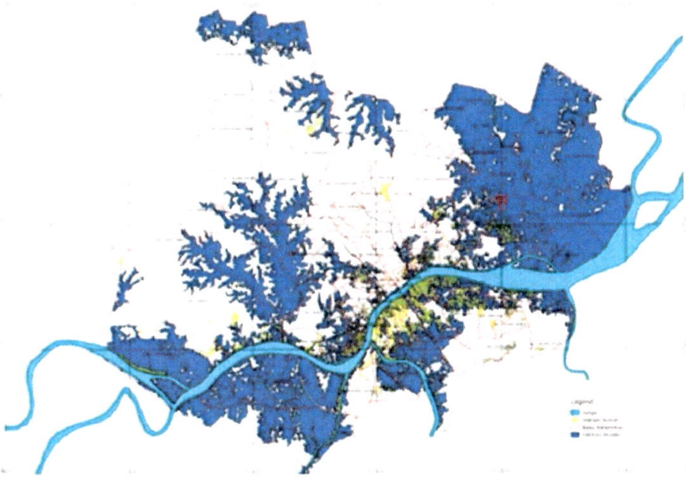

Fig. 3 Slum area overlay map—high water level of Palembang

consisting of permanent and semi-permanent buildings. In addition, the environment is less organized and the lack of infrastructure services such as environmental conditions that are not supported by the area drainage system, waste management system, the absence of a good sanitation system in every house, and damaged environmental roads.

When viewed from the current general description of slums in Palembang, the general classification of slum levels is at the level of mild slums, with a total score of 30 numerical data (range 15–35; Mild Slum). With the main problems in the area, namely the unavailability of fire protection facilities, 98%, unmanaged drainage 86%, and building irregularity 78%. This picture can be seen from the graph of the problem as follows (Fig. 4).

3.5 Slum Characteristics Based on Area Typology

Observation of slum characteristics of Palembang city based on slum area typology is seen from the analysis data of settlement areas based on the tide level of Palembang City. This is because, from the definition that is understood, related to water areas, especially residential areas above water, it is an area that is located above the water level which is affected by tides.

a. Slum Areas on Water, When viewed from the general description of slum areas above water in the city of Palembang, the general classification of slum levels is Mild Slum, with a total numerical data score of 31 (range 15–35; Mild Slum). With the main problems in the area, namely the unavailability of fire protection facilities at 97%, 87% unattended drainage, and 82% irregularity of buildings.
b. Slum areas on the water's edge, when viewed from the general description of slum areas on the water's edge of the city of Palembang at this time, the general classification of slum levels is at the level of Mild Slum, with a total numerical data score of 20 (range 15–35; Mild Slum). With the main problems in the area,

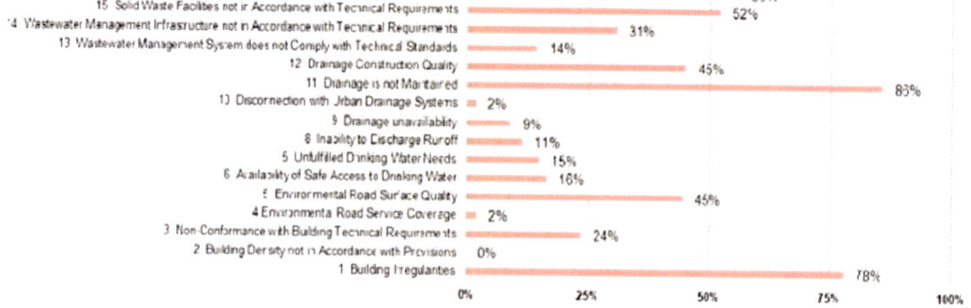

Fig. 4 Graph of slum problems in Palembang (City)

namely the unavailability of fire protection facilities 100.00%, unmanaged drainage 68.00%, and building irregularity 72.00%.

c. Slum Areas in the Lowlands, when viewed from the general description of slum areas in the lowlands of Palembang city today, the general classification of slum levels is at the level of Mild Slum, with a total numerical data score of 26 (range 15–35; Mild Slum). With the main problems in the area, namely the unavailability of fire protection facilities at 95%, not maintaining drainage at 86.00%, and building irregularities at 72.00%.

4 Conclusions

From this research, it can be concluded that the typology and characteristics of slum settlements in Palembang city, namely:

The typology of slum areas in Palembang City is not only in the lowlands and on the water's edge (as stated in the SIAP document), but there are 3 (three) typologies of settlement areas, namely (1) Slum areas on water, (2) Slum area on the water's edge, and (3) Slum settlement area in the lowlands.

In general, the characteristics of slum settlements in the city of Palembang have different trends in the main problems in the area of each typology. For dominant problems with a value of >75%: (1) Slum areas on the water are dominated by the irregularity of the buildings, unattended drainage, and the unavailability of fire protection facilities, (2) Slum areas on the water's edge tend to only have 1 (one) main problem, namely the unavailability of fire protection facilities in the area, (3) Meanwhile, slum areas in the lowlands have 2 (two) problems. The main factors are not maintained drainage system, and unavailability of fire protection system.

Acknowledgements We would like to thank Sriwijaya University for funding this research activity with DIPA UNSRI funds in accordance with the Sateks research contract number: 0179.209/UN9/SB3.LP2M.PT/2018. Apart from that, we would also like to thank Mr. Andi Prasetyo, ST as the KOTAKU Program Assistant Palembang City which has assisted in completing this research. And to the ICSDEMS 2020 board who has facilitated the publication of this report.

References

1. Mardhani H (2012) Penanganan Kawasan Permukiman Kumuh Tepi Sungai Barito Puruk Cahu. Perspektif Arsitektur, 24–34
2. Fitria N, Setiawan R (2014) Identifikasi Karakteristik Lingkungan Permukiman Kumuh di kelurahan Kapuk, Jakarta Barat. Teknik Pomits, 240–244

3. Hariyanto A (2017) Strategi Penanganan Kawasan Kumuh Sebagai Upaya Menciptakan Lingkungan Perumahan dan Permukiman Yang Sehat. (Contoh Kasus: Kota Pangkalpinang). Jurnal PWK Unisba.
4. Nursyahbani R, Pigawati B (2015) Kajian Karakteristik Kawasan Permukiman Kumuh Di Kampung Kota (Studi Kasus: Kampung Gandekan Semarang). Teknik PWK, 267–281
5. Oktarini MF (2015) The house types of settlements in the riverbank wetland; the factual conditions and the recommendations. Tesa Arsitektur:106–116
6. Peraturan Menteri Pekerjaan Umum dan Perumahan Rakyat nomor 2 Tahun 2016 tentang Peningkatan Kualitas Terhadap Perumahan Kumuh dan Permukiman Kumuh
7. Kementerian Pekerjaan Umum dan Perumahan Rakyat (2016) Panduan Penyusunan Rencana Pencegahan dan Peningkatan Kualitas Permukiman Kumuh Perkotaan. PUPR, Jakarta
8. Sori ND (2012) Identifying and classifying slum development stages from spatial data. Geo-information science and earth observation. University of Twente, Netherlands

Food Crisis Transformation to Sustainable Urban Agriculture in Cuba: Lessons for Indonesia

Catharina Any Sulistyowati and Renny Nurhasana

Abstract Urban agriculture has provided many benefits to improve the quality of life, economy, and ecology. Despite its broad benefits, the implementation of urban agriculture in Indonesia is still relatively small compared to its full potential. This paper describes how the severe crisis in Cuba during the Special Period of Peace Time had forced the country to adopt sustainable urban agriculture as one of its policies for achieving food security. The absence of fossil fuel and chemical inputs during the crisis had pushed the country to practice organic agriculture. It used organic fertilizers, biological pest control, local technology, conservation of local seed, and produced for local consumption. Actors and factors are analyzed to understand the context. Based on that, it identified lessons for urban agriculture improvement in Indonesia. Literature review from previous research and publications used as references.

Keywords Sustainable agriculture · Urban agriculture · Food security · Food crisis · Food policy · Agriculture policy · Cuba · Indonesia

1 Introduction

The mainstream development model recognizes the terms center and periphery. This development model has separated the cities and the villages, the urban and rural areas. Cities are centers that generate economic growth. These villages are the periphery that produces food and supports the needs of the cities. Cities are associated with industry and progress, while the villages are associated with agriculture and underdevelopment. These villages produce agricultural products to meet urban dwellers' consumption, while cities produce goods and services to obtain economic

C. A. Sulistyowati (✉) · R. Nurhasana
School of Strategic and Global Studies, Universitas Indonesia, Jakarta, Indonesia
e-mail: catharina.any@ui.ac.id

R. Nurhasana
e-mail: rennynurhasana@ui.ac.id

L. Yola et al. (eds.), *Sustainable Architecture and Building Environment*, Lecture Notes in Civil Engineering 161,
https://doi.org/10.1007/978-981-16-2329-5_18

growth. This development model works because of fossil fuels and various infrastructures that allow the transfer of goods and services between villages and cities [1].

Under this development model, it is easier and cheaper for cities to buy food from the villages than to produce it themselves in urban areas. Part of the reason is that the city's price of land and labor, the main-resources for agriculture, are higher than the price in the village. It created the city highly dependent on food production from its periphery while the periphery relies on the money from the city to fulfill its various needs. It depends on the periphery's food production capacity and food transportation to the center [1: 4–5].

This system can be problematic in the future. Firstly, the increasing population and standard of living in the center require more peripheries to support the center. Forest and water catchment area is converted into farm estate, industrial parks, or settlement to accommodate those needs. It created environmental degradation and reducing the ecological service such as water and material cycles. The impact can be in the form of various disasters such as floods and landslides. If this happens more frequently, it will threaten human security [2].

This system can work only with fossil fuel use that emits carbon dioxide and some other greenhouse gases (GHGs). Both land-use change and fossil fuel use are contributors to climate change. Climate change disaster is happening in an increasing frequency which threatens human life and livelihood. These impacts are experienced differently by age, ethnicity, gender, wealth, and class [3]. Climate change can influence food production in terms of climate uncertainty, in terms of changing temperature and rainfall patterns, which increases the probability of crop failure, which will affect food security in the long run. A study on climate data from 1971–2006 and its impact on food production estimated that Indonesia would have a rice deficit of 90 million tons in 2050. It will be a treat for Indonesian food security, which staple food is rice [4].

Another implication of the system is the packaging use that creates waste problems both in cities and rural areas. The purpose of packaging is to make the products safe during the transfer from the periphery to the center. It also makes the products look more attractive to buyers. After use, the packaging will be thrown away as waste. Improper waste treatment will create pollution in water, land, and marine areas. As the packaging uses durable materials such as plastic, it is not compostable easily by nature. It required a very long time to be composted. The chemical compounds in the packaging color are toxic materials that will enter the water cycles and reduce the quality of drink water [5].

Urban agriculture potentially becomes one of the potential solutions to address the above problems. It will increase city resilience to food crisis as well as reduce the ecological footprint in food production. The implementation of urban agriculture during the crisis is not new. History records that urban agriculture had existed since Ancient Egypt. In the fifteenth century, the Incas had planted vegetables in Machu Picchu—Peru. In the nineteenth century, poor-communities in Germany did urban agriculture as part of their survival strategies. During the first and second

World Wars, urban agriculture was a strategy to provide food supply during war-time [6: 30].

Urban agriculture is an agricultural practice implemented in the city. It can be small-scale intensive agriculture, household food production, land sharing, flower garden, rooftop garden, school gardens, or vertical gardens [7]. "It includes commercial operations that produce food in greenhouses and on open spaces, but is more often small-scale and scattered around the city" [8]. Currently, there are a lot of urban agriculture initiatives in many cities in the world. However, its implementation is still scattered, limited, and far from integration with the national food security strategy. In the global health and ecological crisis, it is crucial to reorganize the food systems to be more sustainable and resilient. Urban agriculture can be one of the strategies.

Cuba had experienced a severe crisis due to the collapse of the Soviet bloc during the nineties. That crisis made its previous agriculture systems that rely on fossil fuel, chemical inputs, and done in rural areas to support urban areas cannot work anymore. That crisis had made Cuba into a hunger condition. Without chemicals, they need to produce food organically. Without fuel, food production and consumption should be local. In short, the lack of those resources had pushed Cuba to make a significant transformation in its food production and consumption into a more local and sustainable one. This model also increases resilience to crisis and promote direct support among community members [9–13].

Cuban experience in implementing sustainable urban agriculture for transforming crisis is extraordinary. Despite so many differences between Cuba and Indonesia in political, social, and economic conditions, they still share at least three similarities. Firstly, Cuba and Indonesia are tropical countries with similar crops and food. Secondly, as an island, Cuba might be comparable to some of the Indonesian islands. Thirdly, the capital city of Cuba, Havana, has a similar size and population number with many capital cities in many provinces in Indonesia.

In COVID-19 pandemics, people and stuff mobility are restricted or limited. In this situation, the importance of urban agriculture can be more crucial [14–16]. Although this pandemic crisis in Indonesia might be less severe than what Cuba had experienced, Cuba's lessons will be useful as examples to find a more resilient, healthy, and sustainable food system in many Indonesian cities.

This paper describes how the severe crisis in Cuba during the Special Period of Peace Time had forced the country to adopt sustainable agriculture in urban areas as the food security policy. The limiting conditions during the crisis had pushed the country to apply more sustainable agriculture. It used organic fertilizers, biological pest control, local technology, conservation of local seed, and produced for local consumption. It explores supporting factors and actors involved in the practices to provide a complete understanding of the context. This understanding will help us deal with current and future crisis, which pushed us to reorganize our food system into more sustainable and localized. It compares Cuban and Indonesian conditions to explore factors to improve urban agriculture in Indonesia. References used literature review from previous research and publications.

2 Crisis and the Transformation to Sustainable Urban Agriculture in Cuba

In Cuba, the growing trend of urban agriculture started after the demise of the Soviet Union in 1989. The tightened US embargo to Cuba through the Torricelli Act in 1992, and the implementation of the Helms-Burton Act in 1996 that gave sanctions to third world countries for doing business with Cuba and restricted the sales of food and medicine to Cuba had put the country into a severe crisis. It resulted in the collapse of previous Cuban agriculture that relied on high technology using fossil fuel imported from the Soviet Union. The impact was the decline of Cuba's sugar and tobacco export and its food security. Without fossil fuel, oxen replaced tractors, while biological pesticides and compost replaced chemical pesticides and fertilizers. Limited fossil fuel also restricted transportation so that cities could not rely on their food from rural areas any more. The urban area had to produce its food. Urban agriculture started to bloom. It became a central element of the country's development [17].

The Cuban government responded to this crisis by implementing two major policy shifts, i.e. (1) the break-up of large-scale state farms into cooperatives (1993) and (2) the opening of the market for agriculture products (1994) [17]. In Cuba, most of the land is state-owned. During the crisis, the state distributed land to those who would like to farm. As a tax, they contributed part of their harvest for school, hospital, and daycare's food supply. President Fidel Castro proclaimed that no piece of land should be left uncultivated [12].

When urban agriculture started in Havana, the topsoil was very poor. It consisted of gravels and little organic material. Soil fertility needs to be improved by increasing organic-soil-compounds. Twenty-five composting centers that processed the city's organic waste into organic compost were set-up. Recycling and the use of local resources are two central elements of the new agriculture practice [12].

Before the crisis, Cuba relied on imported chemical inputs from other countries. This chemical pesticide became unavailable during the crisis. CREE (*Centros Reproductores de Entomofagus y Entomopatogenos*/Center for Reproduction of Entomophagus and Entomopathogens) produced biological pesticides to replace the chemical ones. It uses microorganisms such as fungi and bacteria and plants such as tobacco, melia, and neem. [12, 17: 16].

Since there was nearly no urban agriculture practice in Havana City before the crisis, many city dwellers did not have farming skills. There were efforts done by ACTAF (*Asociacion Cubana de Tecnicos Agricolas y Forestales* or Cuban Association of Agriculture and Forest Technicians) and ANAP (*Asociacion Nacional de Agricultores Pequenos* or The National Association of Small Farmers in Cuba) to train people about technical aspects of urban agriculture. The farmer's cooperatives organized the technical and input provision supports [17].

The Department of Urban Planning and the Department of Urban Agriculture developed zoning for urban agriculture. This zoning is needed to avoid potential hazards from agriculture practice, such as pollution from cattle raising. There is also the involvement of the Ministry of Public Health in mitigating health risks [12: 345].

In Cuba, urban agriculture had saved the country from starvation during the crisis. Although urban agriculture could not cover 100% of Havana's food demand, it had increased the city's food resilience. Local production made a lower transportation cost. It made the food price cheaper and more accessible to people. Direct distribution to schools, hospitals, and daycares created less waste, which in many other places often occurred during the distribution process [12].

Cuban government estimated that urban agriculture had contributed to the employment of more than 117 thousand people directly and an additional more than 26 thousand in agriculture-related jobs. The income of people in these sectors is higher than the country's average income [12].

In the beginning, urban agriculture in Cuba was for the country's survival and not for a green movement. However, the survival strategy had brought Cuba into more sustainable agriculture practice. In addition to what had described above, another aspect of urban agriculture was the reforestation program (*Mi Programma Verde*). In 1997, more than five thousand grassroots projects had plant more than 5 million trees [12].

The environmental impact of urban agriculture was very positive. It used fewer fossil fuels, chemical pesticides, and fertilizers, improved water and air quality, utilized organic waste, and conserve local biodiversity [12]. This practice continues until now. Rafael Santiesteban Pozo, the President of ANAP stated that "the contribution of urban agriculture to the Cuban national economy and food security during the COVID-19 pandemic time has been essential" [15]. As part of the national campaign to raised food at home during the pandemic, the Committees for the Defense of the Revolution (CDR) had distributed flyers among its eight million members. "More than 700 thousand gardens and more than 147 thousand suburban farms were involved in the national movement to increase local food production" [14].

3 Comparison of Urban Agriculture Practice in Cuba and Indonesia

This section discusses some similarities and differences in urban agriculture practice in Cuba and Indonesia. From that discussion, it identifies factors to improve urban agriculture in Indonesia.

Similar to Cuba, urban agriculture in Indonesia also brings positive impacts on the people and the environment. If they do it for their consumption, it will increase family nutrition and food security and reduce their food expenses. If the farmers sell the harvest, they will get increasing income while providing food for the community. In addition to that, urban agriculture also provides ecological functions such as increasing green space in the neighborhood [18, 19].

Although the crisis was less severe than in Cuba, in Indonesia, urban agriculture also had been part of the urban-poor survival strategy during the 1997 economic crisis. They used idle land to grow food to get additional income or to reduce their spending [20]. In Indonesia, the benefit of urban agriculture had been recognized by the government, so that urban agriculture was part of the poverty alleviation program. As an example, in 2007, urban agriculture was part of the poverty alleviation program of the government of Surabaya [18].

Such as in Cuba, urban agriculture became more popular in Indonesia. Although the reasons are slightly different, there are some initiatives in many cities to start urban agriculture projects. In 2011, Ridwan Kamil, the City Mayor of Bandung declared "*Indonesia Berkebun*" which had spread urban farming initiatives to 33 cities in Indonesia [21]. Currently, urban agriculture is not merely a survival strategy of the poor but also an action by the Indonesian middle class to promote a green and healthy lifestyle [22].

A study of urban agriculture in six cities in Indonesia (Surabaya, Cirebon, Bandung, Yogyakarta, Pacitan, and Salatiga) [20] found that most urban farmers are poor. Many of them do urban agriculture as a side job, as a source of additional income, or for using their free time. Most of them cultivate in a relatively small size of land in their backyard or unused land. Most of them used conventional farming, while only a small fraction of them used multiple technologies such as hydroponic, vertical agriculture or mixed farming depends on the land size and condition [20]. It planted various crops such as vegetables, fruits, and ornamental plants. It concluded that urban agriculture had helped the poor during the economic crisis by providing food security, employment, increasing income, and greening the area [19, 20].

Unlike urban agriculture in Cuba that was central for the national food security strategy, urban agriculture implementation in Indonesia is still very small compared to its full potency. During the crisis, the Cuban government integrated all policies to support urban agriculture. It focused on how to produce food for local consumption. The government provided technical assistance, financial support, input provision, and land access. In Indonesia, there are no systematic or integrated urban agriculture policies at the local and national levels. The farmers use their capital to start urban farming. The harvest is sold in local markets or for their personal use [20].

Until now, the city dependence on food supply from its periphery is still very high in Indonesia. A study estimated that 97% of Bandung City's food came from its periphery areas and other provinces [22]. It is similar to the condition of Cuba before the crisis. To increase the city's food resilience, Indonesia can maximize urban agriculture potency to produce food for the cities. It will also provide another benefit of increasing employment in the area [19].

In Indonesia, one of the challenges of urban agriculture implementation is access to land. Most of the land used is idle land owned by individuals, private companies, or governments that can be transferred from time to time depends on the owner's needs [6]. In many cities in Indonesia, the land conversion for housing and industry increased over time. For example, if the trend of this land conversion in Semarang continues, the whole rice field in Semarang will change to other purposes in

66 years [23]. In this situation, there is no guarantee for the long-term implementation of urban agriculture.

Different from Indonesia, 80% of the land in Cuba belongs to the state. The government has the authority to allocate land use according to the needs of the country. In Cuba, the state distributed land to farmers for urban agriculture. It was part of the government's land distribution strategy. In return, the farmers have a clear obligation to pay taxes in the form of giving a portion of their harvest to schools, hospitals, and child care [12, 17]. Urban farmers in Indonesia do not have secure access to the land, nor the long-term cultivation right. There are no definite rules regarding tax obligations for farmers who use idle land for urban farming.

Land size is another issue. A type of urban agriculture uses a small yard. This size limitation brings consequences of limited yields, types of crops, and agriculture technology [6]. However, it also provides rooms for technical innovation on how to farm in limited space, such as wall-gardening, verticulture, rooftop gardens, and hydroponics [19].

During the COVID-19 pandemic, some city people in Indonesia had started home gardening. Many of them put their stories on their social media accounts. So far, no systematic effort by the government to spread this initiative throughout the country. Many options for agricultural development and food security in Indonesia might make urban agriculture less priority. While in Cuba, urban agriculture might be the only option for the country's survival during the severe crisis.

4 Conclusions

From Cuban experience, there are several lessons and recommendations for improving urban agriculture in Indonesia as follows. Firstly, decision-makers and practitioners need to understand the benefit of sustainable urban agriculture for food security, economy, and ecology. This understanding will bring about the interest to implement sustainable urban agriculture to more massive policy support or more bottom-up initiatives by communities.

Secondly, an integrated macro-policy framework to support urban agriculture is essential for urban agriculture in Cuba. This integrated macro-policy framework to support sustainable urban agriculture is still missing in Indonesia. It is very crucial for Indonesia's sustainable urban agriculture development in the future.

Macro-policy frameworks should include the following aspects. (1) Clear long term access to land (land use entitlement). (2) The availability of access to production inputs. It can be in the form of products or a credit to finance the spending on production inputs. (3) The availability of technical support for farmers/gardeners. (4) The availability of supporting systems such as market or other exchange/distribution mechanisms. (5) The economic benefit for farmers/gardeners and communities. Hopefully, all of those factors could increase urban agriculture implementation in Indonesia.

Finally, the COVID-19 pandemic situation can be a good opportunity for Indonesia to start spreading the practice of urban agriculture across its cities. It will not only a coping strategy dealing with the crisis but can be a little step to reorganized the food system into a more sustainable and resilient one.

References

1. von Braun J (2007) Rural-urban linkages for growth, employment, and poverty reduction. In: Ethiopian economic association fifth international conference on the Ethiopian economy. United Nations Conference Center, Addis Ababa
2. Rosyidie (2013) Banjir: Fakta dan Dampaknya, serta Pengaruh dari Perubahan Tata Guna Lahan. J Reg City Plann 24(3)
3. Shukla PR, Skea J, Buendia EC, Masson-Delmotte V (2019) Climate change and land: an IPCC special report on climate change, desertification, land degradation, sustainable land management, food security, and greenhouse gas fluxes in terrestrial ecosystems. IPCC
4. Syaukat Y (2011) The impact of climate change on food production and security and its adaptation programs in Indonesia. ISSAAS 17(1):40–51
5. Stana EJ (1994) Packaging and the environment. Nat Resour Environ 9(2):16–19
6. Sastro Y (2013) Pertanian Perkotaan: Peluang, Tantangan, dan Strategi Pengembangan. Buletin Pertanian Perkotaan 1(3):29–36
7. Fauzi AR, Ichniarsyah AN, Agustin H (2016) Pertanian Perkotaan: Urgensi, Peranan, dan Praktik Terbaik. Jurnal Agroteknologi 10(1):49–62
8. Nugent R (1996) Urban agriculture: an oxymoron? In: Shaw J (ed) The state of food and agriculture. FAO, Rome
9. Altieri MA, Companioni N, Canizares K (1999) The greening of the "barrios": urban agriculture for food security in Cuba. Agriculture and Human Values 131–140
10. Ellinger M, Braley S (2010) Urban agriculture in Cuba. Race, Poverty & the Environment 17(2):14–17
11. Funes F, Garcia L, Bourque M, Perez N, Rosset P (2002) Sustainable agriculture and resistance: transforming food production in Cuba. Appropriate Technology 29(2)
12. Novo MG, Murphy C (1999) Urban agriculture in the city of Havana: a popular response to a crisis. City case study Havana
13. Rosset P, Benjamin M (1994) The greening of the revolution: Cuba's experiment with organic agriculture. Ocean Press, Melbourne
14. Carrero Y (2020a) Organic farming mushrooms in Cuba amid COVID-19 pandemic. https://www.xinhuanet.com/english/2020-09/10/c_139358033.htm. Accessed on 17.11.2020
15. Carrero Y (2020b) Urban gardening takes deeper root in Cuba amid pandemic. https://www.xinhuanet.com/english/2020-07/06/c_139190571.htm. Accessed on 17.11.2020
16. Pangestika D (2020) Urbanites find solace in urban farming amid COVID-19 quarantine. https://www.thejakartapost.com/news/2020/04/12/urbanites-find-solace-in-urban-farming-amid-covid-19-quarantine.html. Accessed on 17.11.2020
17. Sulistyowati CA (2001) Breaking the myths: lessons from agricultural policy change in Cuba during the nineties. Master thesis, Institute of Social Studies, The Hague
18. Santoso EB, Widya RR (2014) Gerakan Pertanian Perkotaan dalam Mendukung Kemandirian Masyarakat di Kota Surabaya. In: Conference on National Cities
19. Sedana G (2020) Urban Farming sebagai Pertanian Alternatif dalam Mengatasi Masalah Ekonomi pada Masa dan Pasca Pandemi Covid-19. http://repository.undwi.ac.id/public/files/penelitian/0001126409_Gede-Sedana_UNDWI_Urban-Farming-sebagai-Pertanian-Alternatif-dalam-Mengatasi-Masalah-Ekonomi-pada-Masa-dan-Pasca-Pandemi-Covid-19_86864.pdf. Accessed on 6.12.2020

20. Setiawan B, Rahmi DH (2004) Ketahanan Pangan, Lapangan Kerja dan Keberlanjutan Kota. Warta Penelitian Universitas Gadjah Mada (special edition):24–42
21. Indonesia Berkebun (2020) https://indonesiaberkebun.org/. Accessed 16.11.2020
22. Puriandi F (2013) Proses Perencanaan Kegiatan Pertanian Kota Yang Dilakukan oleh Komunitas Berkebun di Kota Bandung Sebagai Masukan Pengembangan Pertanian Kota di Kawasan Perkotaan. J Reg City Plann 24(3):227–240
23. Hariyanto (2010) Pola dan Intensitas Konversi Lahan Pertanian di Kota Semarang Tahun 2000–2009. Jurnal Geografi FIS Universitas Negeri Semarang 7(1):1–10

Study of Dry Climate During Extreme El-Nino Occurrence for Plantation Commodities in Nangapanda, East Nusa Tenggara

Nadya Paramitha Putri, Halvina Grasela Saiya, Gilang Buditama, and Lin Yola

Abstract El Nino is an anomaly in sea surface temperature in the Pacific Ocean, resulting in dry conditions and reduced rainfall in Indonesia. This anomaly can cause many things, one of which is a drought that affects plants' growth. As a region that depends on agriculture, agriculture productivity in Nangapanda can be threatened by the dry condition derived by El Nino events. This research aims to detect the spatial and temporal of dry areas and analyze its relationship with plantation commodities productivity in Nangapanda. Landsat 5 TM and Landsat 8 OLI imagery data at the year 2009, 2015, and 2019 were used for analyzing the Normalization of Differences Vegetation Index (NDVI) and Tasseled Cap Transformation (TCT). Overlay of the NDVI and TCT will generate dry areas divided into moderate, high, and very high drought classes. Dry areas during 2019 are 2942.46 ha or 15% of the total area of Nangapanda Subdistrict. Mostly, drought areas are located in the agricultural area and shrubs area. The increasing dry area can cause a decrease in the productivity of plantation commodities in Nangapanda.

Keywords Dry area · El Nino · NDVI · Plantation productivity · TCT Wetness

1 Introduction

In an ecosystem, the climate is a factor that can affect the productivity of the ecosystem. It is very dynamic nature makes it an essential factor in various sectors [1, 2]. Regarding the agricultural sector, the climate is a genuine factor but also risky. If the climate is supportive, it can impact crop success in terms of quantity and quality, but if climatic conditions are not supportive, it will impact drought, floods, hurricanes, extreme temperatures, humidity, etc. This unsupportive condi-

N. P. Putri · H. G. Saiya (✉) · G. Buditama
School of Environmental Science, Universitas Indonesia, Jakarta 10430, Indonesia
e-mail: halvina.grasela@ui.ac.id

L. Yola
School of Strategic and Global Studies, Universitas Indonesia, Jakarta 10430, Indonesia

L. Yola et al. (eds.), *Sustainable Architecture and Building Environment*,
Lecture Notes in Civil Engineering 161,
https://doi.org/10.1007/978-981-16-2329-5_19

tion will ultimately lead to crop failure [2, 3]. Moreover, if it occurs on a large scale, crop failure conditions can result in national agricultural productivity instability.

Climate activities that are often considered to impact the agricultural sector negatively are often referred to as climate anomalies [1, 3, 4]. Climate anomalies is a seasonal shift from the normal average conditions of the season that usually occurs. When climate anomalies occur, the rainy season or dry season can come sooner or later than usual. Climate anomalies that occur globally require all countries to make various predictions based on the collected climate data records, aiming to minimize the impact of losses from the agricultural sector and other sectors [2, 3, 5]. There are five factors in the global climate system that influence each other, namely monsoon winds, local topography, tropical cyclones, zonal cycles, and meridional cycles [4–8]. If one of the five factors dominates the other factors, then the climate anomaly phenomenon will occur [6, 8]. One type of climate anomaly in tropical areas such as Indonesia is El Nino [9, 10]. This phenomenon is one of two types of phenomena caused by ENSO (El Niño Southern Oscillation). This anomaly occurs due to differences in sea surface temperature in the central and eastern Pacific Ocean (the Pacific Ocean area between the Central American Continent and the northern ocean area of Papua New Guinea and other nearby Islands), which is higher or lower than the average temperature. optimum [9–12]. El Nino refers to an increase in sea surface temperature from optimum conditions [11, 12]. When an El Nino occurs, areas of Indonesia, Papua New Guinea, and parts of Australia will experience below normal rainfall levels [9, 10]. This condition has implications for the occurrence of long droughts.

The location of East Nusa Tenggara (NTT) as one of the provinces in Indonesia that has the characteristics of dry areas [13–16], will get drier when El Nino is occurring. Its position is right in the area directly affected by the ENSO anomaly [16, 17]. Conditions prone to drought and the threat of El Nino make agricultural productivity in NTT also threatened. Threats to food crop productivity have a significant effect [18], and their impact on national food security has been widely studied. However, what is no less important is the productivity of longevity or plantation crops, which deserve more attention. This impact is not only related to local revenue [19, 20] but is also able to sustainably maintain farmer productivity when food crops experience severe impacts from climate anomalous activities [19–21]. However, more important is to maintain the resilience of plantation productivity, which will directly protect forest systems that can help reduce the impact of climate change on an area's local ecological resilience [21, 22]. Besides, this is also related to exploration efforts to mitigate the preparation of stocks of superior plantation commodity seeds resistant to the impacts of climate change [20, 23].

One of the districts in NTT that is also proclaimed to be the location of modern seed sources for plantation commodities is Ende Regency [24]. Nangapanda District is one of the sub-districts in Ende Regency, which has very high plantation potential. This can be seen in the data on the area of plantations for the commodity of Cocoa (*Theobroma cacao*) and Coconut (*Cocos nucifera*), which is the widest in Ende District in Nangapanda District [25]. Nangapanda even has many farmer groups that focus on dealing with cocoa [19]. This condition impacts the yields of

Cocoa and Coconut in Nangapanda District, which is the highest of other sub-districts in Ende Regency [25]. Besides these two commodities, it is also followed by Candlenut (*Aleurites moluccanus*), where Nangapanda District has the second-largest land area for candlenut in Ende Regency after Detukeli District [25]. However, the amount of candlenut harvested from Nangapanda almost matches the amount of harvest from Detukeli and does not show a significant difference [25]. It was then followed by the commodity cloves (*Syzygium aromaticum*), where Nangapanda is one of the top ten producers of cloves in Ende Regency based on the calculation of annual productivity results [25]. With these various superior commodities, it is essential to study the climatic factors that affect plantation yields in Nangapanda.

It was considering the extreme drought conditions that occurred in 2009, 2015, and 2019. Based on Oceanic Nino Index (ONI) data, it can be seen that in 2009 there was an increase in temperature that exceeded the optimum conditions for six consecutive months, so that drought conditions were created [26]. In 2015, these conditions occurred throughout the year, which caused a full year of drought [26], and finally, in 2019, the drought phase occurred within eight months when there was a fluctuation in temperature increase above the optimum range [26]. These years can be categorized as extreme El Nino years. Based on these things, the purpose of this paper is to see the effect of extreme El Nino conditions on the yield of plantation commodities in the Nangapanda District. This is important to study by looking at the distribution of dry areas when these extreme conditions occurred, namely 2009, 2015, and 2019. After analyzing this, the analysis will be developed to see the real effect on superior commodities' productivity—plantations in Nangapanda, namely Cocoa, Coconut, Candlenut, and Cloves. A remote sensing approach using Normalized Difference Vegetation Index (NDVI) and Tasseled Cap Transformation (TCT) Wetness is used to achieve this goal. NDVI and TCT Wetness are spatial analysis techniques that can provide correlative visualization results by the conditions that occur [27–30]. This method's apparent advantage is that it can complement and provide a basis for more in-depth statistical evidence. For this reason, the index obtained from the analysis is combined with a regression test to see the significant effect on the yield of plantation commodity productivity.

2 Methods

2.1 Study Area

This research was conducted at the end of 2019 to early 2020 in Nangapanda District, Ende Regency, East Nusa Tenggara Province, with the coordinates of the location at 08° 39′ 12″–08° 48′ 35″ S and 121° 23′ 39″–121° 34′ 26″ E (Fig. 2), about 35 km to the west of Ende City. Results of calculations using remote sensing stated that the area of Nangapanda District is about 190.3196 km^2, and bordered by several surrounding areas. At the north and west, respectively, it is bordered by Maukaro District and Nagekeo District, while at the south and east, respectively, it is bordered by the Sawu Sea and Ende District (Fig. 2). The field observations indicate that this area is a mostly dry land with a combination of mountain topography, hills, valleys, and beaches. The climatic conditions in Nangapanda are classified as dry, and this is based on the results of recording the intensity of rainfall, which show that in a year, there are eight dry months and four wet months [31].

Administratively, there are 28 villages in Nangapanda [31], with most people making ends meet by farming. They take advantage of the dry land conditions in Nangapanda by maintaining the forest while planting and developing longevity plants for plantations and food crops suitable for dryland conditions. Also, a small part of the wetland agricultural potential is being developed in Nangapanda, namely lowland rice. Wetland farming is specially developed in villages that are very close to the potential of abundant water sources so that throughout the year, their agricultural land continues to be well watered. However, most of the villages in Nangapanda, which were far from access to water sources, were found in most villages, so agricultural and food crops suitable for dry areas were developed in these villages.

2.2 Data

The data used in this research are Landsat 8 OLI/TIRS and Landsat 5 TM multi-temporal imagery data and the data of primary plantation productivity in Nangapanda from Agricultural Instructor Agency. Three series of Landsat imageries were acquired from the United States Geological Survey (USGS). In this case, the data series used in 2009, 2015, and 2019 was recorded in the drought season, especially from April to October. Land use maps from Geospatial Agency, soil type maps from the Soil Research Institute, and average annual rainfall from 2009–2019 obtained from three measurement stations support the result of this research.

2.3 Vegetation Density Index

The vegetation density was obtained from Landsat imageries using the calculation of Normalized Difference Vegetation Index (NDVI). The vegetation density index is one way to sharpen the contrast between vegetation and soil surface [32, 33]. According to Tucker [34], the formula of NDVI is:

$$NDVI = \frac{Band\ NIR - Band\ Red}{Band\ NIR + Band\ Red}$$

The calculation results from the NDVI will produce a value of −1 to 1. Areas with an index value of NDVI below 0.1–0.2 are considered the non-vegetation group (because they can be water areas or rocky soil). Areas with NDVI values above 0.4 can be concluded that the area is covered with vegetation and is fertile but is still classified as moderate vegetation density. Meanwhile, for NDVI value above 0.6, it can have a high vegetation density [35–37] (Table 1).

2.4 TCT Wetness

Tasseled Cap Transformation (TCT) is a mathematical formula for calculating the level of brightness, greenness, and Wetness from digital numbers of each band (band 1 to band 5 and band 7) on Landsat imagery [38]. TCT was first introduced by Kauth and Thomas [39] for Landsat MSS. Furthermore, TCT was refined by Crist and Cicone [40] using Landsat TM data. The three values in TCT can be used to analyze soil dryness, especially with a feature named Wetness, which was defined to correspond to the observed direction of soil moisture variation in the plane occupied by bare soils data [41–43].

According to Hasan et al. [44], the formula of TCT Wetness for Landsat 8 OLI/TIRS is:

$$\begin{aligned} Wetness =& (0.1511 \times Band\ 2) + (0.1973 \times Band\ 3) + (0.3283 \times Band\ 4) \\ &+ (0.3407 \times Band\ 5) - (0.7117 \times Band\ 6) - (0.4559 \times Band\ 7) \end{aligned}$$

According to Crist and Cicone [40], the formula of TCT Wetness for Landsat 5 TM is:

$$\begin{aligned} Wetness =& (0.1509 \times Band\ 1) + (0.19731 \times Band\ 2) + (0.3279 \times Band\ 3) \\ &+ (0.3406 \times Band\ 4) - (0.7112 \times Band\ 5) - (0.4572 \times Band\ 7) \end{aligned}$$

Table 1 NDVI value based on previous research

Class	Classification	NDVI Value
1	Non vegetation	<0.2
2	Low density vegetation	0.2–0.4
3	Moderate density vegetation	0.4–0.6
4	High density vegetation	0.6–0.8
5	Very high density vegetation	>0.8

The negative soil moisture index shows a low soil moisture level, while soil moisture with positive values indicates higher soil moisture [38]. Meanwhile, the TCT Wetness value used in this research is (Table 2).

2.5 Distribution of Dry Area

The assessment of the dry area level was carried out by integrating the greenness index and the soil moisture index. The use of greenness and soil moisture levels has been carried out in several previous studies [45–47]. This integration is carried out because each index's value cannot be used to determine drought conditions, where the results of the NDVI or TCT analysis have the same appearance in different drought conditions so that there can be a misinterpretation. It is necessary to integrate by combining greenish values (NDVI) and soil surface moisture (TCT). The integration of greenness and soil surface moisture can be done using a matrix (Table 3).

The assessment of the dry area level was carried out by integrating the level of greenness (NDVI) and the level of soil moisture (TCT Wetness). The use of greenness and soil moisture levels for assessing the dry area has been carried out in several previous studies [45–47]. This integration is carried out because the value of each index cannot be used to determine dry area, where the results of the NDVI or TCT analysis may have the same appearance in a different dry area so that there can be misinterpretation and it is necessary to integrate by combining greenness values (NDVI) with the soil surface moisture (TCT). The integration of greenness and soil surface moisture can be done using a matrix (Table 3).

2.6 Data Analysis

The data analysis used in this research is spatial analysis and statistics. The spatial analysis was performed by overlapping the NDVI data with TCT Wetness data to obtain the distribution of dry areas in Nangapanda. Statistical analysis was performed by testing and described the relationship between the dry areas with agricultural productivity.

Table 2 TCT Wetness value (Data processing, 2020)

Class	Classification	TCT Wetness value
1	Very high (very moist)	0.60–0.38
2	High (moist)	0.38–0.16
3	Moderate	0.16–(−0.06)
4	Low (dry)	(−0.06)–(−0.28)
5	Very low (very dry)	(−0.28)–(−0.5)

Table 3 Matrix of dry area assessment [38]

Class of dry area	Soil moisture level					Vegetation density level
	Very low (1)	Low (2)	Moderate (3)	High (4)	Very high (5)	
Very dry (severe)	1.1	1.2	1.3	1.4	1.5	Non vegetation (1)
	2.1	2.2	2.3	2.4	2.5	Low density vegetation (2)
Dry	3.1	3.2	3.3	3.4	3.5	Moderate density vegetation (3)
Moderate dry	4.1	4.2	4.3	4.4	4.5	High density vegetation (4)
Non dry area	5.1	5.2	5.3	5.4	5.5	Very high density vegetation (5)

3 Results and Discussions

3.1 *Distribution of Dry Area*

The dry area in Nangapanda tends to increase, from 2009 with a totally dry area of 1226.16 ha to 2015 with a totally dry area of 1596.96 ha, and 2019 with a totally dry area 2942.46 ha. The increase in totally dry area from 2009 to 2019 is more than doubled, 58.3% (Figs. 1 and 2). The increase in a dry area is due to a decrease in the NDVI index value, which means there is the degradation of vegetation (for example, from shrubs that resembles forests into shrubs that is not too dense; or from forests with high index value into the plantation area with lower index value) or change of land use from vegetated to non-vegetated. This increasing dry area is also due to a decrease in the TCT Wetness value, which states that there is a change in soil moisture level due to the change in land use.

Spatially, the dry area dominates the southern part of Nangapanda, which is a beach and hilly ridge. In this southern part, land use in dry areas is dominated by agricultural areas, grasslands, and shrubs. Meanwhile, the rest is distributed in the

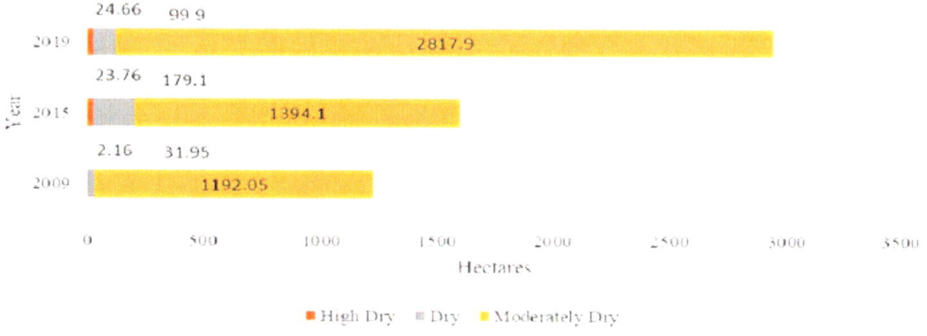

Fig. 1 Dry area in Nangapanda

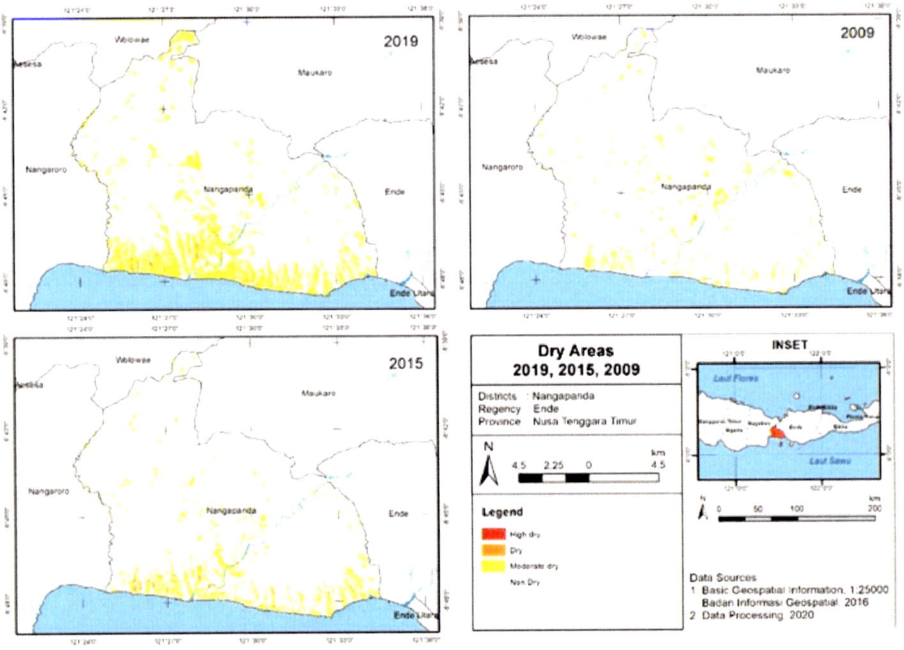

Fig. 2 Distribution of dry areas in Nangapanda 2019, 2015, 2009

central and northern parts of land in agricultural areas and shrubs. Temporally, the distribution of dry areas occupies the same area, but the dryness gets worse over time.

3.2 Correlation of Dry Area and Productivity

The farmers in Nangapanda use their land for one type of agricultural commodity, but the farming tradition passed down from generation to generation has also created a mixed farming system in Nangapanda. In one land ownership, most of the land will be allocated for plantation commodities, and the rest is for food crops. So there is no single farmer in Nangapanda who only develops plantation crops or only food crops. All of them use the same system as their ancestors did. The choice of plant types also adjusted to the estimated seasons that will occur in a year. However, specifically for plantation crops, Nangapanda chose the types of commodities to be planted based on government programs' socialization and priority through the Ministry of Agriculture. So that by paying attention to the climate and the conditions for plant growth, five plantation commodities have been developed, namely cocoa, coconut, candlenut, coffee, and cloves. Cocoa, coconut, and candlenut continue to survive and experience increased production but are inversely related to coffee. Furthermore, the last commodity that has experienced promising development in Nangapanda is cloves. For this reason, this paper focuses on data for cocoa, coconut, candlenut, coffee and cloves.

The productivity of the primary plantation commodities in Nangapanda is fluctuating (Table 4). Cocoa and coffee tend to decline in the study area, especially cocoa, which experienced a decrease up to 0.4 ton/ha from 2009 to 2019. The candlenut and coconut also experienced a decrease in productivity but not significantly. Meanwhile, the productivity of cloves is increasing.

The El Nino events can cause dry conditions and a reduction in rainfall that can affect plants' growth. Based on the interviews with the officer of Nangapanda Agricultural Instructor Agency and farmers in Nangapanda, there was a season anomaly where the rainy season in 2014 and 2019 experienced a delay. Based on the correlation results, only cloves have a positive relationship with dry areas. This suggests that cloves' productivity increases with the increase of dry area, or it can be said that the productivity of clove is not disturbed by an increase in a dry area. Meanwhile, cocoa, candlenut, coffee, and coconut have a negative relationship with dry areas. This implies that the productivity of these commodities will decrease along with the increase of dry areas, or it can be said that the productivity of these commodities will be affected by an increase in dry areas (Table 5).

Table 4 Productivity of plantation commodities (tonnes/ha)

Year	Dry area	Productivity of plantation commodities (tonnes/ha)				
		Coconut	Cocoa	Candlenut	Coffee	Cloves
2009	1226.16	1.075	1.000	1.000	0.500	0.300
2015	1596.96	1.201	0.988	0.899	0.511	0.333
2019	2942.46	1.000	0.630	1.000	0.326	0.400

Source Nangapanda Agricultural Instructor Agency [31]

Table 5 Correlation of dry area and plantation commodities

Correlations							
		Dry_Area	Coconut	Cocoa	Candlenut	Coffee	Cloves
Pearson correlation	Dry_Area	1.000	−0.744	−0.984	−0.076	−0.966	0.992
	Coconut	−0.744	1.000	0.850	−0.610	0.890	−0.656
	Cocoa	−0.984	0.850	1.000	−0.102	0.997	−0.955
	Candlenut	−0.076	−0.610	−0.102	1.000	−0.183	−0.198
	Coffee	−0.966	0.890	0.997	−0.183	1.000	−0.928
	Cloves	0.992	−0.656	−0.955	−0.198	−0.928	1.000
Sig. (1-tailed)	Dry_Area		0.233	0.057	0.476	0.083	0.039
	Coconut	0.233		0.176	0.291	0.150	0.272
	Cocoa	0.057	0.176		0.467	0.026	0.096
	Candlenut	0.476	0.291	0.467		0.442	0.437
	Coffee	0.083	0.150	0.026	0.442		0.122
	Cloves	0.039	0.272	0.096	0.437	0.122	

Referring to the correlation and spatial test results, it is found that the El Nino phenomenon affects plantations in Nangapanda. El Nino phenomenon from year to year can be represented by the area of dry areas, and plantation activities can be represented by the productivity of plantation commodities. With the results of this research, it is expected that the community, farmers, government, and related agencies can develop agricultural strategies to anticipate the impact of drought due to the El Nino phenomenon on plantation productivity. As well as being able to build resilience in the region if there is a decrease in plantation productivity or even crop failure.

4 Conclusions

Based on the data analysis results, the dry area in Nangapanda has doubled in the last ten years, from 2009 to 2019. The dry area distribution dominates the southern part of the Nangapanda area, consisting of beaches and hilly ridges and occupied by land use of agricultural areas and shrubs. Dry areas correlate with the productivity of plantation commodities in Nangapanda, where cocoa, coconut, candlenut, and coffee commodities have decreased along with the increase in dry areas. Meanwhile, cloves did not experience a decline despite an increase in dry areas.

This data needs to be considered by local governments and plantation owners in Nangapanda to build local resilience to deal with the impact of El-Nino anomalies, especially for plantation productivity. The selection of superior seeds and types of plants that are resistant to the impacts of climate change needs to be considered carefully. With this, it is hoped that local governments and communities will not only select plantation commodity seeds to be planted based only on predictions of economic benefits. But it is also very important to pay attention to their resilience to extreme climatic conditions.

Acknowledgements This publication were supported by DRPM (Directorate of Research and Community Service) Universitas Indonesia through PUTI Research Grant with contract number NKB-1271/UN2.RST/HKP.05.00/2020. We want to send our gratitude to the people of Nangapanda. Furthermore, our big thanks to Sanca Pamungkas for helping us to collect data in Nangapanda. Moreover, our gratitude to the Center for Human Resources and Environmental Research (PPSML), that facilitates us to conduct an initial survey through cooperation funds with contract numbers 358G/UN2.F4.D2.2/LT- KEP.SPK/X/2019. An assistance was also provided by the DIVERLING (Biodiversity for Environmental Sustainability) Research Cluster, School of Environmental Science, Universitas Indonesia. This paper's map can also be done with the help from the Geospatial Laboratory ("*Laboratorium Geospasial SIL*") at the School of Environmental Sciences, through using the Map Info Pro license software, ArcGIS ESRI license, and ENVI.

References

1. Ayanlade A, Radeny M, Akin-Onigbinde AI (2018) Climate variability/change and attitude to adaptation technologies: a pilot study among selected rural farmers' communities in Nigeria. GeoJournal 83(2):319–331
2. Alidoost F, Su Z, Stein A (2019) Evaluating the effects of climate extremes on crop yield, production and price using multivariate distributions: a new copula application. Weather Clim Extremes 26:100227
3. Heusser LE, Hendy IL, Barron JA (2015) Vegetation response to southern California drought during the medieval climate anomaly and early little ice age (AD 800–1600). Quatern Int 387:23–35
4. Xie SP, Kosaka Y, Du Y, Hu K, Chowdary JS, Huang G (2016) Indo-western Pacific ocean capacitor and coherent climate anomalies in post-ENSO summer: a review. Adv Atmos Sci 33 (4):411–432
5. Yang K, Zhang J, Wu L, Wei J (2019) Prediction of summer hot extremes over the middle and lower reaches of the Yangtze River valley. Clim Dyn 52(5–6):2943–2957
6. Hsu HH, Zhou T, Matsumoto J (2014) East Asian, Indochina and western North Pacific summer monsoon-an update. Asia-Pacific J Atmos Sci 50(1):45–68
7. Fine CM, Johnson RH, Ciesielski PE, Taft RK (2016) The role of topographically induced vortices in tropical cyclone formation over the Indian Ocean. Mon Weather Rev 144 (12):4827–4847
8. Poan ED, Gachon P, Dueymes G, Diaconescu E, Laprise R, Sanda IS (2016) West African monsoon intraseasonal activity and its daily precipitation indices in regional climate models: diagnostics and challenges. Clim Dyn 47(9–10):3113–3140
9. Qian JH, Robertson AW, Moron V (2010) Interactions among ENSO, the monsoon, and diurnal cycle in rainfall variability over Java, Indonesia. J Atmos Sci 67(11):3509–3524
10. Alsepan G, Minobe S (2020) Relations between interannual variability of regional-scale Indonesian precipitation and large-scale climate modes during 1960–2007. J Clim 33 (12):5271–5291
11. Rashid HA, Hirst AC (2016) Investigating the mechanisms of seasonal ENSO phase locking bias in the ACCESS coupled model. Clim Dyn 46(3–4):1075–1090
12. Roy I, Kripalani RH (2019) The role of natural factors (part 1): addressing on mechanism of different types of ENSO, related teleconnections and solar influence. Theoret Appl Climatol 137(1–2):469–480
13. Jayanti AD, Fitriya W, Setyobudi E, Budhiyanti SA, Kune SJ (2018) The vulnerability of fishermen's community and livelihood opportunity through drought and seasonal changes in border area of Indonesia-Timor Leste. E&ES 139(1):012030
14. Kuswanto H, Fithriasari K, Inas R (2018) Drought risk mapping in East Nusa Tenggara Indonesia based on return periods. Asian J Sci Res 11(4):489–497
15. Kuswanto H, Hibatullah F, Soedjono ES, Efendi F (2019) Survey data of household perceptions of drought, mitigation and adaptation practices in East Nusa Tenggara, Indonesia. Data in brief 24:103944
16. Patandean CF (2020) Influence of Enso to Variability of Rain Chain at Tropical Cyclon in Southern East Nusa Tenggara. Jurnal Ilmiah Ilmu Administrasi Publik 9(2):277–286
17. Hornidge AK, Scholtes F (2011) Climate change and everyday life in Toineke Village, West Timor: uncertainties, knowledge and adaptation. Sociologus: 151–175
18. Lobell DB, Asseng S (2017) Comparing estimates of climate change impacts from process-based and statistical crop models. Environ Res Lett 12(1):015001
19. Tola D (2016) Analisis Produksi Pertanian Kakao Di Kecamatan Nangapanda Kabupaten Ende. Jurnal Pendidikan Ekonomi, Ekspektasi 1(1):26–39
20. Lillesø JPB, Harwood C, Derero A, Graudal L, Roshetko JM, Kindt R, Moestrup S, Omondi WO, Holtne N, Mbora A, van Breugel P (2018) Why institutional environments for agroforestry seed systems matter. Dev Policy Rev 36:O89–O112

21. Calfapietra C, Gielen B, Karnosky D, Ceulemans R, Mugnozza GS (2010) Response and potential of agroforestry crops under global change. Environ Pollut 158(4):1095–1104
22. Pawson SM, Brin A, Brockerhoff EG, Lamb D, Payn TW, Paquette A, Parrotta JA (2013) Plantation forests, climate change and biodiversity. Biodivers Conserv 22(5):1203–1227
23. Ghini R, Bettiol W, Hamada E (2011) Diseases in tropical and plantation crops as affected by climate changes: current knowledge and perspectives. Plant Pathol 60(1):122–132
24. Directorate of Plantation, Ministry of Agriculture (2020) https://www.pertanian.go.id/home/index.php?show=repo&fileNum=251
25. Central Bureau of Statistics of Ende Regency (2019) Ende regency in numbers. Ende Regency
26. NOAA (2020) Cold & warm episodes by season. https://origin.cpc.ncep.noaa.gov/products/analysis_monitoring/ensostuff/ONI_v5.php
27. Hogg EH, Barr AG, Black TA (2013) A simple soil moisture index for representing multi-year drought impacts on aspen productivity in the western Canadian interior. Agric For Meteorol 178:173–182
28. Bobkov A, Panidi E, Torlopova N, Tsepelev V (2015) NDVI dynamics of the Taiga zone in connection with modern climate changes. In: International archives of the photogrammetry, remote sensing & spatial information sciences
29. Chandrasekar K, Sai MS (2015) Monitoring of late-season agricultural drought in cotton-growing districts of Andhra Pradesh state, India, using vegetation, water and soil moisture indices. Nat Hazards 75(2):1023–1046
30. Rossi M, Niedrist G, Asam S, Tonon G, Tomelleri E, Zebisch M (2019) A comparison of the signal from diverse optical sensors for monitoring alpine grassland dynamics. Remote Sens 11 (3):296
31. Nangapanda Agricultural Instructor Agency (2019) Programa Penyuluhan Pertanian 2019. BPP Nangapanda, Nangapanda
32. Rouse JW, Haas RH, Schell JA, Deering DW (1974) Monitoring vegetation systems in the Great Plains with ERTS. In: Third earth resources technology satellite–1 symposium, vol 1, pp 309–317
33. Gitelson AA, Kaufman YJ, Merzlyak MN (1996) Use of a green channel in remote sensing of global vegetation from EOS-MODIS. Remote Sens Environ 58:25–31
34. Tucker CJ (1979) Red and photographic infrared linear combinations for monitoring vegetation. Remote Sens Environ 8:127–150
35. Sudiana D, Diasmara E (2008) Analisis Indeks Vegetasi menggunakan Data Satelit NOAA/AVHRR dan TERRA/AQUA-MODIS. In: Seminar on intelligent technology and its applications
36. Sobrino JA, Jiménez-Muñoz JC, Paolini L (2004) Land surface temperature retrieval from landsat TM5. Remote Sens Environ 90:434–440
37. Taufik A, Ahmad SSS, Ahmad A (2016) Classification of landsat 8 satellite data using NDVI thresholds. J Telecommun Electron Comput Eng 8(4)
38. Shofiyati R, Kuncoro D (2007) Remote sensing for drought assessment on agricultural land. Informatika Pertanian 16(1):923–936
39. Kauth RJ, Thomas GS (1976) The tasselled cap—a graphic description of the spectral-temporal development of agricultural crops as Seen by LANDSAT. LARS Symposia. No. 159
40. Crist EP, Cicone RC (1984) A-physically-based transformation of thematic mapper data—The TM tasseled cap. IEEE Trans Geosci Remote Sens GE-22(3):256–263
41. Rahman S, Mesev V (2019) Change vector analysis, tasseled cap, and NDVI-NDMI for measuring land use/cover changes caused by a sudden short-term severe drought: 2011 Texas event. Remote Sens 11(19):2217
42. Susilowati MHD, Wibowo A, Susiloningtyas D, Rastuti RC (2019) Mapping of drought and food insecurity region in Lebak regency Banten Province. GeoEco J 5(2):127–140
43. Crist EP, Kauth RJ (1986) The tasseled cap de-mystified. Photogram Eng Remote Sens 52 (1):81–86

44. Hasan ABM, Zhang L, Shuai T, Tong Q (2014) Derivation of a tasselled cap transformation based on landsat 8 at-satellite reflectance. Remote Sens Lett 5(5):423–431
45. Wahyunto H, Suryani E et al (2016) Petunjuk Teknis Pedoman Penilaian Kesesuaian Lahan untuk Komoditas Pertanian Strategis Tingkat Semi Detail Skala 1:50.000. Bogor: Balai Besar Penelitian dan Pengembangan Sumberdaya Lahan Pertanian, Badan Penelitian dan Pengembangan Pertanian, 37 pages
46. Saha A, Patil M, Goyal VC, Rathore DS (2019) Assessment and impact of soil moisture index in agricultural drought estimation using remote sensing and GIS technique. In: Proceedings MDPI 7(2)
47. Anugerah MM, Wibowo A, Kusratmoko E (2019) Detection of drought area using greening index and wetness index in Sukaresmi Subdistrict, Cianjur Regency, West Java. IOP Conf Ser: Earth Environ Sci:311

The Choice of Transportation Mode in Transit Location (Case Study: Dukuh Atas Area)

Roby Dwiputra and Lita Sari Barus

Abstract Traffic in Jakarta is one of the main issues that becomes an annual problem in the capital. According to World Bank data, Jakartans spend at least 3.5 h in traffic jams on average [1]. Being one of the main causes, among others, there are more than 18.6 million private modes of transportation in Jakarta, while users of public transportation only reach 24% [2]. In supporting the choice of public transportation modes to reduce the traffic ratio in Jakarta, it is not sufficient by only providing adequate public transportation, but also requires a city development concept that provides convenience and comfort for road users, including zoning, passenger flow and intermodal integration. Recent studies attempted to find evidence of a relevance between TOD and transit location. The area around Dukuh Atas Station which is planned to be the center of the development of one of the TODs in Jakarta becomes a transit location for 4 modes of public transportation. However, it is difficult for passengers to get on and off at this transit location to move from one mode of public transportation to another, so that passengers choose the private mode of transportation to continue their trip to their destination. Preferences Survey done by using a survey questionnaire conducted in this study determine the preferences of passengers so that they still choose to use public transportation. The results of the study are known factors that passengers desired in order to switch to public transport in Dukuh Atas transit locations which also provide suggestions to the Government and the management of public transport modes to develop this area to be oriented to the comfort for the passengers.

Keywords Dukuh Atas · Passenger · Choice of transportation mode · Transportation public · Preference survey

R. Dwiputra · L. S. Barus (✉)
Urban Studies, School of Strategic and Global Studies, University of Indonesia, Jakarta, Indonesia
e-mail: litabarus@ui.ac.id

L. Yola et al. (eds.), *Sustainable Architecture and Building Environment*, Lecture Notes in Civil Engineering 161,
https://doi.org/10.1007/978-981-16-2329-5_20

1 Introduction

Jakarta as the National Capital City of Indonesia is a city with a high population density and a high and increasing rate of urbanization. In line with this, traffic is one of the main problems in Jakarta that still cannot be resolved. Concept of urban development that provides convenience and comfort for road users is needed, including areaal arrangement, the flow of passengers and intermodal integration in addition to the provision of adequate public transportation to support the election of public transport modes by passengers. In making the concept of urban development that provides convenience and comfort for road users, the Government plans to develop area of TOD in Dukuh Atas area that carry the vision of the development which is the collaboration of motion in space of the city, improve the quality the life of urban communities through urban movement infrastructure improvements towards a humanist urban space. One of the principles of this development is increased connectivity, namely a simple, direct and intuitive transit connection.

The problems in Dukuh Atas area is the facility unavailability in facilitating the activities of transit and the incapability to make passengers remain comfortable to continue to use public transport which causes passengers tend to switch to private transport for the trip. It can be concluded that at the Dukuh Atas transit location, public transportation is unable to compete with private transportation. If this situation is left abandoned, the users of public transportation will continue to decrease. It is necessary to know the preferences of passengers to continue choosing public transportation at transit locations in the Dukuh Atas area in continuing the trip to the next destination. Therefore, this research is important to maintain sustainable public transportation, where the results of this study can also provide suggestions for the Government, both central and local, managers of public transportation modes, and managers of the Dukuh Atas Area to develop this area with a convenience-oriented orientation to the users.

2 Methods

2.1 Framework

This research started from a theoretical review and then collected research data. Analyzing then were conducted using quantitative descriptive statistical analysis to describe the data that has been collected (Fig. 1).

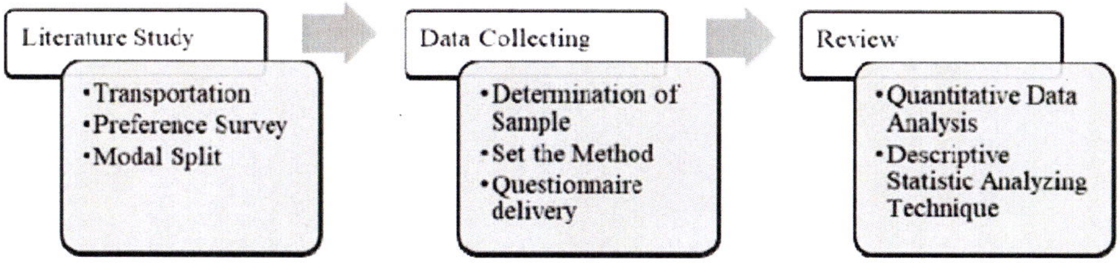

Fig. 1 Framework research

2.2 *Theory Overview*

Transportation is the movement of people and/or goods from one place to another that arises because of activities in the community. According to Tamin [3], the classification of transportation types can be divided into two, namely:

1. Public transportation, is a mode of transportation that is provided for the needs of the movement of people for the sake of public interests or the public in fulfilling their needs; and
2. Private transportation is a mode of transportation which in its operation can freely determine its own line or route as long as it does not violate traffic regulations.

Level of service mode of transportation public is the quality and quantity provided by the mode manager. The variables measuring the level of public transportation services according to Agustin [3] include travel time, fare, accessibility, waiting time, comfort and safety. These variables consist of quantifiable variables such as travel time and rates as well as variables that are difficult to quantify such as accessibility, waiting time, comfort and safety.

Modal Split or modal choice is a modeling or stages of the planning process transport that serves to determine the loading per street or to know the number of people and goods as users to choose various modes of transport available to serve a point of origin-specific purposes, and also for certain trip purpose. Factors that may affect the modal split can be grouped into three, that is characteristic of the road, movement characteristics and characteristic modes of transport facilities [4] (Table 1).

Table 1 Affecting factors of modal split

Factor	Description
Road users	These are vehicle ownership, driver's license ownership, household structure, income, and the necessity to use other vehicles to work
Movement	It is the purpose of the movement, the time when the movement occurs and the distance traveled
Facility	These include travel time, costs, space availability and estimated rates, as well as comfort, safety and reliability

Fig. 2 Dukuh Atas area and transit location

Transit Oriented Development, hereinafter abbreviated as TOD, is an area development concept based and centered on mass public transportation stations, which accommodates new growth into a mixed area with an area of 350–700 m from the center of the area which is integrated with the area surroundings through the utilization of ground, lay and underground space. The TOD area is a mixed residential and commercial area with high accessibility to mass public transportation, where mass public transport stations and mass public transport terminals are the center of the area with high density buildings. Meanwhile, mass public transportation is public transportation that can carry large numbers of passengers, which operates fast, safe, secure, scheduled and high frequency (Fig. 2).

Dukuh Atas area which is the case study in this research is located in 3 (three) districts and 2 (two) administrative city areas, namely Menteng District and Tanah Abang District, Central Jakarta Administrative City, and Setiabudi District, South Jakarta Administrative City. The area around Dukuh Atas Station which is planned to be the center of the development of one of the TODs in Jakarta will become a transit location for 4 modes of public transportation, including MRT, CL, BRT and Airport Train. Apart from the four existing transit locations, this area is planned to be the location for LRT transit.

2.3 *Preference Survey*

Survey is the process of collecting and measuring data which has at least 3 (three) characteristics, namely: (1) Aim to produce statistics which are quantitative or numerical descriptions of several aspects of the population under study; (2) Collecting information by asking structured questions to get answers and process the answers into data analysis; and (3) The information collected is a fraction of the population called the sample.

Consumer preferences can be interpreted as a taste of preference, choice or something that consumers like [5]. The purpose of the preference survey in this study is to determine the preferences and needs of respondents who are passengers of public transportation modes in the Dukuh Atas area, including those who transit at the Transjakarta Tosari BRT Stop, Sudirman CL Station, Dukuh Atas MRT Station, and BNI City Airport Train Station.

2.4 Determine of Sample

The sample is part of the population studied and becomes a description for the population in the study [6]. The population in this study is the total number of passengers during 2019 which passes through the station/bus stop public transport mass transit in Dukuh Atas area. Based on data obtained from various sources, the number of CL passengers on the Depok/Bogor—Jatinegara route is 73.84 million people [7], Transjakarta BRT corridor 1 is 30.530.769 people [8], the Airport Railway is about 2 million people [9], and the MRT is 19.990.950 people [10]. So that the total number of passengers as the population size is 126.361.719 people.

After obtaining the population size, then the writer uses the Slovin formula to determine the minimum sample size with a significance level/magin error of 10% with the following calculation results:

$$n = \frac{N}{1 + Ne^2} = \frac{126.361.719}{1 + 126.361.719(10\%)^2} = \frac{126.361.719}{1.263.618,19} \approx 100$$

Legend:

n minimum number of sample size
N total population size
e level of significance/margin of error.

Then this research uses the *Nonprobability Sampling* method, this method does not provide equal opportunities for every member of the population to be selected as a sample. The sampling technique used the *Reliance on Available Subjects* [11], sampling based on available subjects, namely who are members of the population who can be contacted by the author can be a data source and used as a sample.

2.5 *Descriptive Statistical Analysis*

Descriptive statistical analysis is a part of statistics that studies how to collect data and present data so that it is easy to understand. Descriptive statistics are only related to describing or providing information about a data or situation. With the word descriptive statistics function to explain the condition or problem [12].

3 Result and Discussions

The research discussion is in the form of data analysis using descriptive statistical analysis techniques by describing the statistical presentation of the data that has been collected. The number of samples studied based on respondents obtained from distributing questionnaires within 5 (five) days is 110 respondents.

3.1 *Questionnaire Result*

Based on the results of the questionnaire, we can see that the transit place most frequently visited by passengers is the CL Sudirman Station chosen by 76 respondents, then followed by the Transjakarta Tosari BRT Stop chosen by 36 respondents, then Dukuh Atas MRT Station was chosen by 26 respondents and finally the BNI City Airport Train Station chosen by 16 respondents. This also illustrates that the largest number of mass public transport passengers transiting in the Dukuh Atas area are CL passengers (Fig. 3).

The results of this questionnaire indicate that there are still 54% of respondents who still choose to switch to private transportation when transiting in the Dukuh Atas area. While 24 respondents (22%) answered maybe yes/maybe no, most answered this depending on weather conditions (Fig. 4).

Fig. 3 Transit location respondents

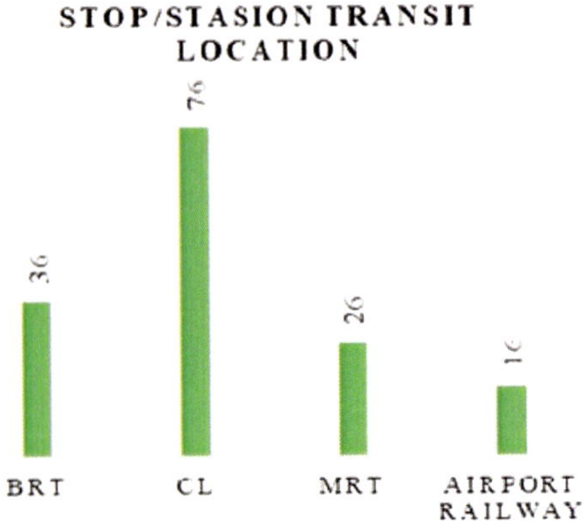

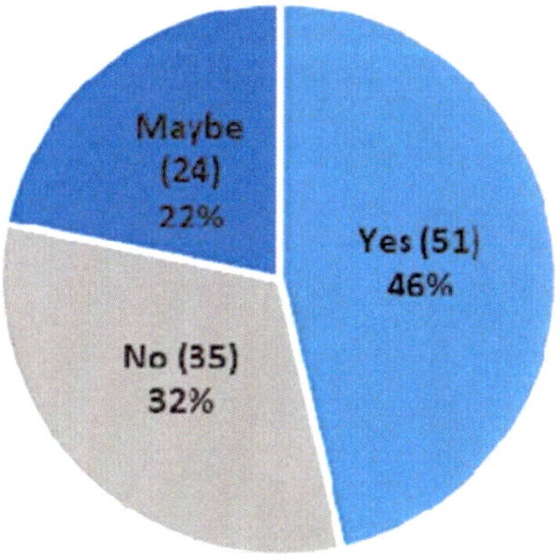

Fig. 4 Respondents willingness to go on with public transportation

3.2 Public Transportation Service Level

The level of service (Fig. 5) provided by public transportation modes is a determining factor for passengers in choosing public transportation. Based on the results of the questionnaire, the transportation mode service that affects respondents is dominated by factors of travel time, waiting time, convenience and accessibility. Meanwhile, the respondent considered the safety factor, cost and other factors to be good so that the effect did not have a significant impact on the choice of public transportation modes by passengers.

3.3 Passenger Preferences

There are 3 (three) main things that passengers want in the integration of public transportation modes (Fig. 6). First, the movement of passengers from one transit point to another is safe and comfortable, especially for the elderly, children and

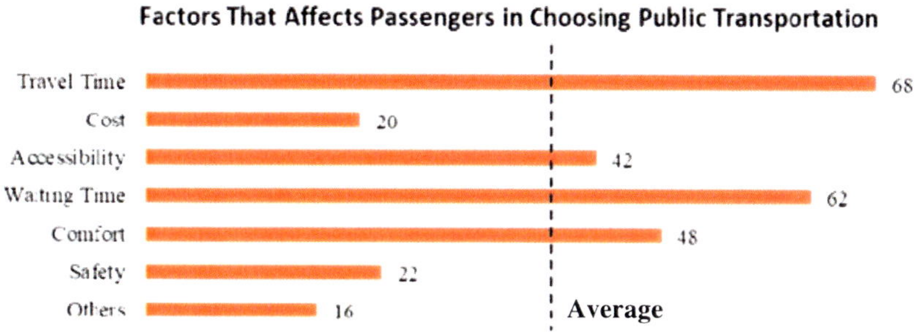

Fig. 5 Questionnaire results of the factors in choosing public transportation

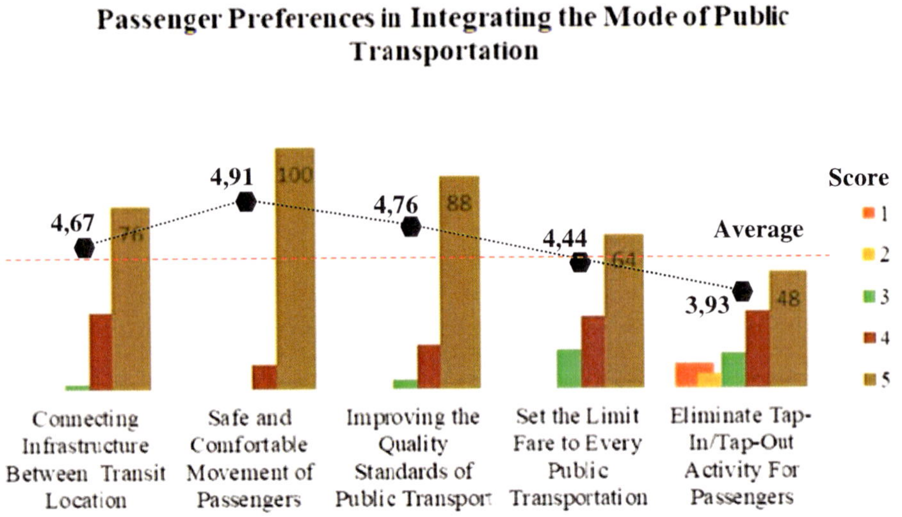

Fig. 6 Questionnaire results of preference respondents in integrating public transportation

pregnant women. Second, equalization and improvement of quality standards for public transportation modes so that all modes of public transportation have a balanced level of comfort and performance. Then the third is the development of infrastructure in form of connecting facilities between bus stops/stations (Table 2).

That the activity of equalizing the quality of the transportation mode standards has the greatest significance by increasing the 5 (five) factors of the size of public transportation services. This was followed by the construction of connecting infrastructure between transit points and ensuring safe and comfortable movement of passengers, each of which increased 4 (four) factors. Then the elimination in tap-in/tap-out activities which increased 3 (three) factors and finally the fare

Table 2 The results of the matrix analysis of the relationship between the size of transportation service size variables and passenger preferences, 2020

Factor	Connecting infrastructure transit location	Safe and comfortable movement of passengers	Improving the standards of public transport	Set the limit fare to every public transportation	Eliminate tap-in/ tap-out activity for passengers
Time travel	■		■		■
Cost fees			■	■	
Accessibility	■	■			■
Waiting time		■	■		■
Comfort	■	■	■	■	
Safety	■	■	■		

restrictions on each public transportation mode was followed by 2 (two) factors that increased.

4 Conclusions

The author makes conclusions based on the results of the research discussion and also provides suggestions to the Government and Public Transportation Mode Managers in integrating modes in the Dukuh Atas transit location. The conclusions of this study include:

1. The results showed that 54% of respondents were still unwilling/or unsure about choosing public transportation. Switching mode to private transportation are caused by public transportation were outplayed in intermodal competition in Dukuh Atas transit location. The variables of public transportation services that dominate in influencing passenger choices are travel time, waiting time, convenience and accessibility
2. Managers of public transportation modes to improve service level, especially the punctuality of travel time: arrival, departure and passengers transit as well as increasing comfort both in public transportation modes and in transit.
3. The Government together with managers of public transportation modes to encourage an equalization and improvement of quality standards for each mode of transportation so that all modes of public transportation have a balanced level of comfort and performance.
4. The Government should build infrastructure in form of connecting facilities between bus stops/stations, then together with managers of public transportation modes to ensure that the movement of passengers from one transit point to another is safe and comfortable, especially for the elderly, children and pregnant women.

References

1. Danang Sugianto (2017) Detik Finance at https://finance.detik.com/berita-ekonomi-bisnis/d-3641306/warga-jabodetabek-kena-macet-35-jam-tiap-hari-kotanya-rugi-rp-39-t
2. Anonym (2019) MRT Jakarta at https://www.jakartamrt.co.id/konektivitas/transit-oriented-development-tod
3. Agustin T (2006) Analisis Variabel Layanan Angkutan Umum. UNS, Surakarta
4. Tamin OZ (2000) Perencanaan dan Pemodelan Transportasi. ITB, Bandung
5. Case & Fair (2007) Jasa Perdagangan. Kanisius, Yogyakarta
6. Martadiputra BAP (2012) Populasi dan Sampel. UPI, Bandung
7. Puspa, Anita Widya (2020) Bisnis.com at https://bisnis.tempo.co/read/1302224/beroperasi-dua-tahun-kereta-bandara-soetta-sepi-peminat/full&view=ok
8. Sandiputra R (2020) Berita Jakarta at http://www.beritajakarta.id/read/76168/jumlah-penumpang-transjakarta-2019-meningkat-signifikan

9. Cahyani DR (2020) Tempo.co at https://bisnis.tempo.co/read/1302224/beroperasi-dua-tahun-kereta-bandara-soetta-sepi-peminat/full&view=ok
10. Anonym (2019) Media Indonesia at https://mediaindonesia.com/read/detail/275168-8-bulan-beroperasi-penumpang-mrt-20-juta-orang
11. Babbie E (2012) The practice of social research, 13th edn. Wadsworth, Cangage Learning, Medina
12. Hasan I (2001) Pokok-pokok Materi Statistik 1 (Statistik Deskriptif). PT. Bumi Aksara, Jakarta

The Impact and Prevention of the Pandemic Crisis on the Sustainability of Private Universities in Indonesia

David Ronald Tairas and Tuntas Widyo Soenanto

Abstract Indonesia for the first time confirmed the Covid-19 case in March 2020 announced by President Joko Widodo, the impact of the pandemic made the economy unstable which also affected higher education by decreasing the number of students enrolling, so leaders, especially private higher education, took several strategies In order to survive the pandemic, this study took a sample of 69 respondents from private universities in 34 provinces in Indonesia, the research method used quantitative methods using the SPSS 25 measuring instrument, it was found that the Low Cost strategy was the most widely used strategy, leaders of private universities in Indonesia.

Keywords Impact · Prevention · Pandemic · Sustainability

1 Introduction

The Pandemic that has occurred in the country since last March has had an impact on various sectors, including higher education. To prevent the spread of the virus from spreading in tertiary institutions, the Ministry of Education and Culture from the beginning has been alert to respond to this condition by issuing Circular Number 3 of 2020 concerning Prevention of Covid-19 in the Education Unit. As a result, universities, both public and private, were forced to eliminate teaching and learning activities on campus and switch them to distance learning ("Media Indonesia", May 18, 2020). The new order due to the Pandemic made students adapt to the online system implemented by the campus, students were satisfied with the assistance provided by the campus administration staff, all students, including lecturers, implemented a healthy life using masks and washing hands to avoid

D. R. Tairas (✉)
School of Strategic and Global Studies, Universitas Indonesia, Jakarta, Indonesia
e-mail: david.ronald@ui.ac.id

T. W. Soenanto
STIE-Tri Dharma Widya, Jakarta, Indonesia

L. Yola et al. (eds.), *Sustainable Architecture and Building Environment*, Lecture Notes in Civil Engineering 161,
https://doi.org/10.1007/978-981-16-2329-5_20

Covid-19. Aristovnik et al. [1]. Chairman of the Indonesian Private Higher Education Association, Budi Djatmiko said 70% of students in private universities (PTS) are in arrears for single tuition fees or educational development contributions. According to him, the Covid-19 pandemic has made it difficult for students from the lower middle class to pay ("Republika", August 09, 2020). Indonesia currently has around 4200 universities and colleges. Among them, only 85 are public universities. The rest, or the majority, are private universities, the majority of which are small campuses. The number of students is less than 1000 people ("Ekonomi Bisnis", June 16, 2020). Based on the description in the introduction of the research question is how private colleges can survive the pandemic crisis.

2 Literature Review

According to Michael E. Porter [2], if a company wants to increase its business in increasingly fierce competition, the company must choose business principles, namely high-priced products or low-cost products, not both. Based on these principles, Porter stated that there are 3 Generic strategic principles, namely the Low Cost Leadership Strategy, Differentiation, and Focus.

1. Low Cost Leadership Strategy (overall cost leadership). The characteristic is that the company concentrates its attention on low product selling prices to reduce production, promotion and research costs. If necessary, the resulting product simply imitates the design product
2. Differentiation Strategy (differentiation). The characteristic is that the company decides to build a potential market perception of a superior product/service so that it looks different from competitors' products. Customers are expected to be willing to buy at a high price because of the differences.
3. Strategy Focus (focus). The characteristic is that the company concentrates on a certain market share to avoid competitors.

In his research D.R. Tairas, Low Cost has a significant effect on the competitive advantage of private universities in Jakarta. Tairas and Mahlia Muis [3]. To implement a cost leadership strategy, corporations need to pay attention to the skills and resources of corporations and corporate organizations. The skills and resources required are: continuous capital investment and access to capital, process engineering skills, strict labor supervision, products designed for ease of production and low cost distribution systems. Kodrat [4], in order to attract consumer attention and create market share, new players (newentry) on generally use the lowcost strategy. Companies with a costleadership strategy are required to implement ability to design, manufacture, and offer its products efficiently compared to competitors (Rusliati [5]: 68). The method used to execute costleadership with production costs, promotion, and research is relatively low. The advantage of this strategy lies in sensitive buyers against price or using price as a determining factor for decisions

(Ardhan and Hatane [6]: 59). Product innovation is proven to be able to mobilize the market and achieve competitive advantage. Kuncoro W. [7], "*Outcomes of competitive advantages dominated sources of advantages in predicting future firm profitability persistence*". Maury [8], "*There is a strong correlation between the level of exercising excellence strategies in education and the achievement of higher educational institutions to the sustainable competitive advantage*". MJ Al Shobaki et al. [9]

According to Kartajaya [10], differentiation can be built based on several aspects or dimensions, namely: The differentiation dimension refers to the "what" value the company offers to consumers. This is the tangible part of differentiation. Content differentiation as the main offering from products and companies to customers. Context Is a dimension that refers to the "way" (how to offer) the company offers value to customers. Where companies differentiate themselves from competitors based on how to offer value to customers. Infrastructure Enabling factors for the realization of content differentiation, the context above, where this dimension shows the differences against competitors based on technology capabilities (technology), HR capabilities (people) and ownership of facilities to support creating content differentiation and the above context.

3 Methodology

Quantitative research is a statistical analysis, Measure objective facts, Focus on variables, Reliability the critical factor, Value-free, Separate theory and data, Independent of context study of the parts and phenomena, and the causality of their relationships to social security and electoral statistics, and other related databases [11]. The objective of quantitative analysis is to develop and use mathematical models, theories, and hypotheses related to natural wonders. The measurement process is a central part of quantitative research because it provides a fundamental relationship between empirical observations and quantitative relationships' mathematical expression [12].

3.1 Steps in the Quantitative Research Process by W. Lawrence Neuman

Figure 1 steps in the Quantitative Research Process; it explains how the researcher must choose the topic to be studied. The second step is the Focus Question, and the research is limited by not widening the question. The third step is the Design Study; the researcher makes the research framework than; the fourth step is to collect the data by doing After surveying respondents. In the fifth step, the data obtained are processed and analyzed according to the number of respondents. Published so that other researchers can get information about the research.

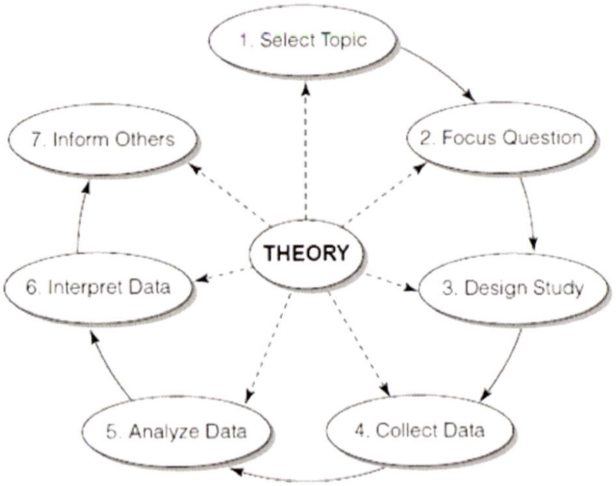

Fig. 1 Steps in the quantitative research process by W. Lawrence Neuman

3.2 Research Framework

In Fig. 2: Research Framework variable Low-Cost Leadership Strategy (X1), Variable Differentiation Strategy (X2) and Strategy Focus (X3) are Independent Variables, while the Sustainability Variable is the Dependent Variable, in the Research Framework using Theory of Michael E Poter (1980) In this study, there are three hypotheses. The first is H1 there is a significant relationship between the independent variable X1 Low-cost leadership strategy with the dependent variable Y. H2; there is a positive relationship between the X2 Differentiation Strategy variable with the Y Sustainability H3 variable; Strategy Focus (X3) has a significant effect on Bound Variables or Sustainability.

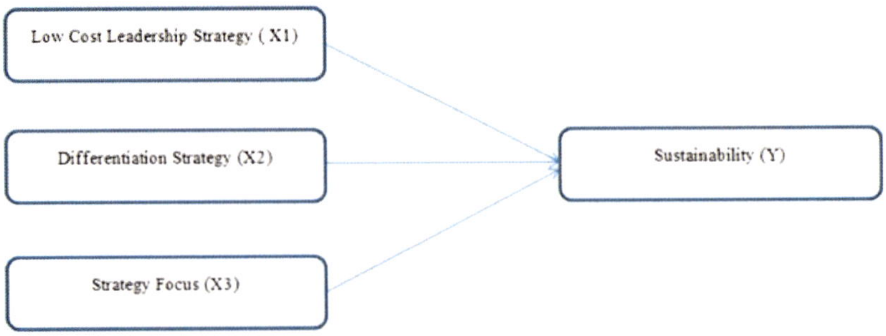

Fig. 2 Research framework

Table 1 Profile of respondents

Characteristics	Percentage %
Gender	
Male	68
Female	32
Total	100
Level of education	
Bachelor degree	51
Master	29
Doctor	20
Total	100
Age (years old)	
20–30	30
31–40	50
51–60	20
Total	100
Length of work	
1–5 years	40
6–10 years	25
>10 Years	35
Total	100

3.3 Research Data Sources

Sources of research data were obtained from 34 private tertiary institutions in 34 provinces in Indonesia, 69 respondents' data consisting of 47 men and 22 women, Of a total of 69 respondents, there are 68% men and 32% women. Men still dominate the management of higher education. In terms of education, 51% of them are Bachelor's Degree, and 29% have a Master's degree. In contrast, 20% are Doctor Education. that there is no equality of education, while there are 40% of the length of workers who work for less than five years, and 25% for 6 to 10 years of work, while 35% for working periods over ten years (Table 1).

4 Partial Regression Coefficient

From the results of calculations using SPSS. T test is used to determine the effect of partially independent variables (Low Cost Leadership Strategy, Differentiation Strategy and Strategi Focus) on the dependent variable (Sustainability), below is presented Table 2.

By the explanation in Table 2, it can be stated that the Variable X1 Low Cost Leadership Strategy with a significant value of 0.000, thus it can be said that (X1)

Table 2 T test (partial)

Variables	Regression coefficient	T-test	Sig.
(Constant)	0.515	1.997	0.547
Low cost leadership strategy	0.491	3.466	0.000
Differentiation strategy	0.405	3.176	0.001
Strategy focus	0.108	2.005	0.003

has a positive effect on the dependent variable (Y) Sustainability with a percentage value of 49.1% while the Variable X2 Differentiation has a significant value of 0.001 with a percentage of 40.5% a significant effect on Y1 Sustainability and X3 Strategy Focus having a significant value of 0.003 meaning that it affects the dependent variable with a percentage of 10.8% so that Hypotheses 1, 2 and 3 are accepted.

5 Conclusion

Private universities during a pandemic use various strategies in the sustainability of their institutions. From the results of the study, it can be concluded that the public prefers low cost in choosing private universities with diversity in the implementation of strategies carried out by leaders of private universities, helping their institutions survive during the current Pandemic, impact From a pandemic, universities have to adjust to the situation, with the data obtained in this study being input for other universities to use strategies from Michael E Poter, it turns out that this strategy is still relevant to the demands of the Pandemic.

References

1. Aristovnik A, Keržič D, Ravšelj D, Tomaževič N, Umek L (2020) Impacts of the COVID-19 pandemic on life of higher education students: a global perspective. Sustainability 12:8438
2. Porter ME (1980) Competitive strategy: tehniques for analyzing industries and competitors
3. Tairas DR, Mahlia Muis SE (2016) The influence of strategic leadership and dynamic capabilities through entrepreneurship strategy and operational strategy in improving the competitive advantage of private universities in Jakarta, Indonesia. Higher education
4. Kodrat DS (2009) Membangun Strategi "Low Budget High Impact" di Era NeW wave Marketing. Jurnal Manajemen Bisnis 2(1):59–86
5. Rusliati E (2014) Pengaruh Informasi Akuntansi Manajemen terhadap Fungsi Manajemen, Kebijakan Produksi, dan Kinerja Proses Bisnis Internal. J Trikonomika 68–70
6. Ardhan JD, Hatane DSE (2015) Analisa Pengaruh Intellectual Capital dan Inventory Turnover terhadap Profitabilitas Perusahaan. Bus Acc Rev 3(1)
7. Kuncoro M (2003) Metode RisetUntukBisnis Dan Ekonomi. Erlangga, Jakarta
8. Maury B (2018) Sustainable competitive advantage and profitability persistence: Sources versus outcomes for assessing advantage. J Bus Res 84:100–113

9. Al Shobaki MJ, Abu Naser SS, Abu Amuna YM, El Talla SA (2017) Impact of electronic human resources management on the development of electronic educational services in the universities. Int J Eng Inf Sys 1(1):1–19
10. Kartajaya H (2009) Hermawan Kartajaya Seri 9 Elemen Marketing on Service. Markplus & Co
11. Neuman WL (2014) Social research methods: qualitative and quantitative approaches seventh edition
12. Creswell JW (2003) Research design: qualitative, quantitative, and mixed methods approaches. SAGE

Comparative Study of Thermal Performance of Building Wall Materials in Bauchi State Nigeria

Olutobi Gbenga Ayegbusi and Lin Yola

Abstract This study focus on the performance of selected secondary school building walls in the hot and dry tropical Climate of Bauchi, Bauchi state Nigeria. The objective of present study was to explore the most efficient way to retrofit existing public schools in Bauchi as the state government embarks on extensive renovation work of the existing school building to provide a more conducive environment for the student's academic activities. Six different wall configurations including the existing wall were simulated during hot season using Design Builder simulation tool to evaluate the wall impacts on the classroom air temperature and relative humidity (RH). The simulation results analysis shows that although the wall configuration with wall insulation produce a lower indoor air temperature, yet it may not be sufficient to provide the needed and thermal comfort. Therefore, to achieve appropriate level of thermal comfort in Bauchi public schools, emphasis should be on both the walls and the roof. This study shows that improving only the wall thermal performance may be an insufficient strategy.

Keywords Thermal performance · Indoor air quality · Wall materials · School building · Tropical climate

1 Introduction

There has been a growing concern regarding the state of affairs of public secondary school in Nigeria due to lack of maintenance over a protracted long period of time. This has led to unsafe, unhealthy and hazardous environment to learn for the country's coming generation. Many researches have documented the poor state of physical facilities in Nigerian Government Secondary Schools [1–3]. Providing a comfortable learning space and sustainable building in the twenty-first century Nigerian environment can be met if the stakeholders believe in efficient retrofitting

O. G. Ayegbusi · L. Yola (✉)
School of Strategic and Global Studies, Universitas Indonesia, Jakarta, Indonesia
e-mail: lin.yola@ui.ac.id

L. Yola et al. (eds.), *Sustainable Architecture and Building Environment*, Lecture Notes in Civil Engineering 161,
https://doi.org/10.1007/978-981-16-2329-5_20

of these buildings thereby providing a sustainable built environment and hence respond accordingly with necessary policies to achieve the goal [3, 4].

In addition, there are studies on the relationship between Indoor Air Quality (IEQ) and student's academic performance. According to Haverinen-Shaughnessy (2007) [5], there are documentation of poor IEQ on the academic performance of students. High indoor air temperature and inadequate ventilation have been identified as some of the main indoor environmental quality problems [5]. In hot and dry environment like Bauchi, excessive high indoor air temperature and indoor air quality is a challenge. Similarly, thermal comfort is an important variable in the evaluation of indoor environment quality, which is dependent on the facade design, the envelope assembly and construction material types [6, 7].

This study aims at investigating three wall materials and few configurations of wall compositions that consist of these materials and the insulation to explore the performance with respect to the air temperature in the selected classroom.

Providing the required level of comfort in an indoor space is dependent on adequate understanding of the relationship between outdoor and indoor [8] with respect to the indoor comfort parameters. Oral et al. [6] along with many other research observed that indoor environment's design parameters ought to be determined during design phase of a project in tandem with the required comfort level [7]. Unver et al. [9] documented design strategies to achieve optimum envelop, especially where there is the need to balance the design visual, thermal and acoustic performances [7, 9].

Building envelope is made up of transparent and opaque materials. Deringer et al. [10] noted that transmission through envelope include the heat through these components, heat gain from indoor sources such as the lighting features, the occupants and the equipment. Further studies also include the impacts of different building envelope components, the effects of wall thermos-physical properties on the time lag and decrement factor [10].

The performance of different building envelope materials vary depending on the inherent properties of building materials that include the heat transmission heat and absorption properties [11]. Both Jannat [10] and Balaji et al. [12] identified material's thermal conductivity, density, and specific heat capacity are basic inherent thermos-physical properties required for the materials' thermal performance. Vijayalakshmi et al. [13] studied the thermal performance of opaque materials under the influence of solar heat and the resulting impact on the interior enviornment.

2 Methodology

The present research focused on providing optimum building wall design for appropriate thermal comfort in the chosen classroom. Therefore, this section presents the methodology, research assumptions and the data collection strategy from the building site.

The use of simulation tools to explore the thermal performance of building components and the internal heat gains has become popular among designers and researchers. This research methodology is favoured compared to manual calculation for efficiency and time saving. Therefore, this study employed DesignBuilder, which has been proven to provide reliable results through verification and validation results [14–19].

2.1 Research Study Case and Simulation

Bauchi city is located between latitudes 9° 3′ and 12° 3′ North and longitudes 8° 50′ and 11° East on an altitude of 616 m above the sea level. According to Koppen-Geiger, classification of climate zones, the city falls within the tropical (hot and dry) climate zone. The intensity of solar radiation on buildings in similar climate do cause overheating and thermal discomfort in the indoor environment as many related studies have shown [14]. The monthly average air temperature is within the comfortable range except for the month of March through May when the air temperature is above 27.6 °C [20, 21].

The building case selected for this study is Government Day Secondary School, Army Barracks, Bauchi, Bauchi State Nigeria. The experiment data was collected during the hot season in 2018 with the data logger positioned at 1.1 m height above the finished floor level as shown in Fig. 1. The data collected include the air temperature, humidity and air speed. The classroom layout is a single space block with 3.2 m height, covered with an ivory white aluminium-roofing sheet at 25° slope and a 0.6 m roof overhang. The walls are made of hollow sandcrete block materials of 450 mm × 225 mm × 225 mm and the floor is finished with cement/ sandcreed. The 1200 × 1500 mm window is made of steel material, with 1200 × 300 mm fixed.

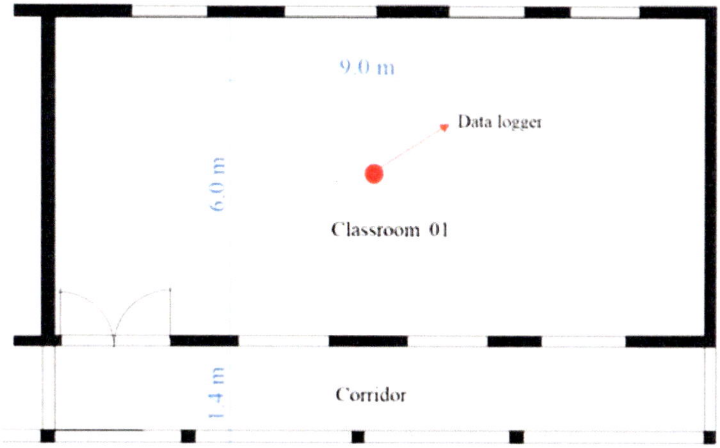

Fig. 1 Building case floor layout with data logger location

The classroom space was modelled with DesignBuilder in stages. At the first stage, the model was simulated with existing wall materials. Afterwards, the wall compositions and configurations were altered to determine the effects on the classroom air temperature and humidity. The sections of six different wall configurations including the existing (W1-W6) simulated are as shown in Fig. 2.

The simulation was run with neither source of internal heat gain nor heating/cooling systems. However, 0.25 Air Changes per Hour of infiltration rate was included to indicate natural air leakage through the envelope of the space and the window set at 100% for natural ventilation. Other parameters are as shown on Table 1.

3 Results and Discussion

The study results show that the classroom air temperature reached the pick at 14:00 h with 41.31 and 38.51 °C for outdoor and indoor respectively. These air temperatures are over the comfort level by more than 10 °C. In contrast, the least outdoor and indoor relative humidity data was 52.67% and 58.06% at mid-day respectively. Both the outdoor and indoor RH data is above 60% during the morning hours and are mostly within comfortable range of 30–60% as given by ASHRAE [21].

Table 2 presents the overall U-value and R-value for different wall configurations choosing for the present study. The classroom air temperature profile presented in Fig. 3 shows that the least recorded value was early in the day and

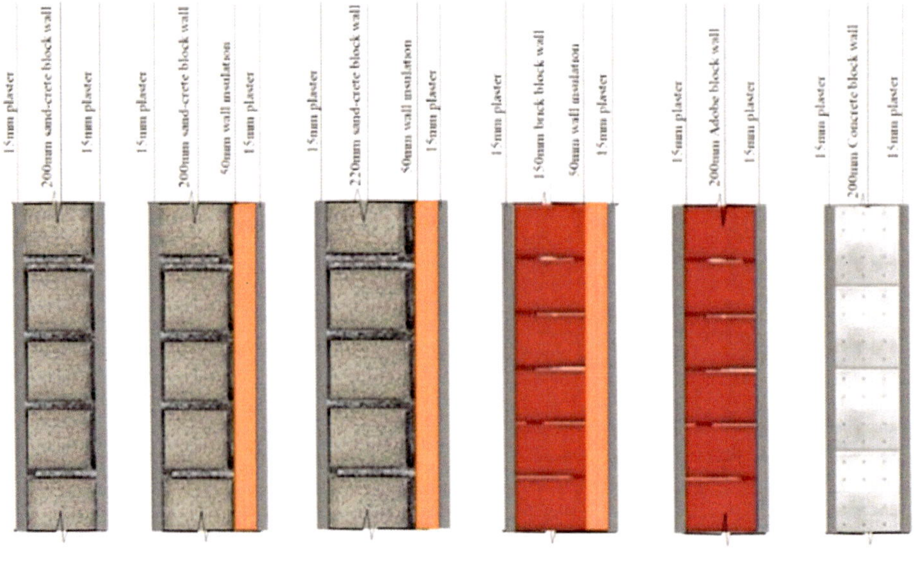

Fig. 2 Different wall construction types

Table. 1 Simulation parameters

Solar distribution	Full interior and exterior
Temperature Control	Air temperature
Surface convection algorithm (in & out)	TARP
Outside convection algorithm	DOE-2
Air mass flow exponent	0.5
Zone capacitance multiplier	50
Time step per hour	6
Sky diffuse modelling algorithm	Simple sky diffuse modelling

Table. 2 Wall data for decision support matrix

Wall configurations	R (W/m-K)	U-value
W1	1.004	0.996
W2	2.514	0.398
W3	2.586	0.387
W4	2.049	0.488
W5	1.849	0.533
W6	0.433	2.311

consistently increased throughout the day. The results also showed that a general increase in the wall thickness did caused a decrease in U-value and hence a lower indoor air temperature. However, a 230 mm thick concrete wall produced a much higher U-value compared to a similar size of brick or masonry wall. The introduction of a 50 mm insulation board on each of the wall configurations (W2, W3 & W4) helped the classroom overall air temperature throughout the day (Fig. 3a).

Figure 3b shows the profile of the calculated relative humidity throughout the day. The results show that the relative humidity from 11:00 am are within 30–60% range, which is considered as a comfortable range according to ASHRAE standard

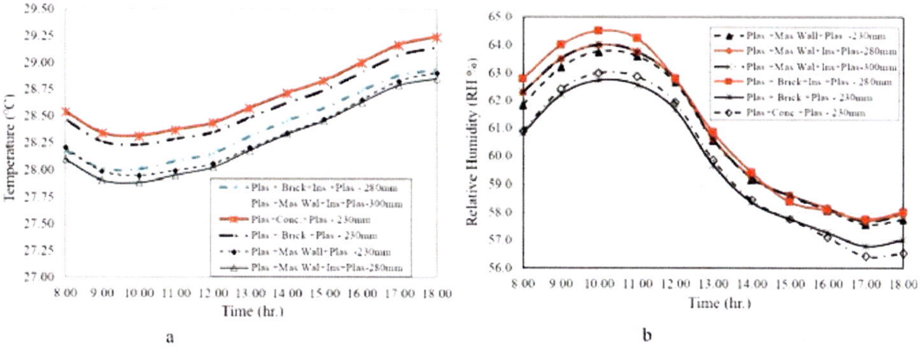

Fig. 3 Indoor **a** air temperature and **b** relative humidity of the classroom with different wall configurations

[21]. At the early hours though, the RH was above 60%, which does not necessarily cause discomfort but could lead to mold growth, mildew and many other indoor environment issues.

4 Conclusions

The present study reinforces the existing fact that the overall building envelope design is important to the thermal comfort of the occupants. The results from this study show that insulating the building walls indeed reduce the overall air temperature of the learning space from 1 to 10 °C throughout the day. However, the air temperature may still be above recommended comfort level if the roof is not refurbished as well. It can be concluded therefore, that both the walls and the roof component should be improved to achieve the required level of comfort in the classroom. It can also be inferred from the study that concrete wall is not recommended as it performs poorly compared to other tested wall configurations. Finally, further work would include exploring the impact of insulated roof on the classroom air temperature and the impact on the overall thermal comfort.

References

1. Yadar K (2001) The teaching of life sciences. Anmol Publication Ltd. India, New Delhi
2. Yara PO, Otieno KO (2010) Teaching/learning resources and academic performance in mathematics in secondary schools in Bondo District of Kenya. Asian Soc Sci (ASS) 6(12):98–115
3. Ojeje MA, Adodo A (2018) Education infrastructure in Nigeria: an analysis of provision of school building facility in secondary schools in Delta State Nigeria. J Edu Entrepreneurship 5(3):49–60. https://doi.org/10.26762/jee.2018.400021
4. Izobo-Martins O (2014) Maintenance strategies and condition of public secondary school buildings in ado-odo/ota local government area Ogun State, Nigeria. Doctoral dissertation, Covenant University, Ota Nigeria
5. Shaughnessy RJ, Haverinen Shaughnessy U, Nevalainen A, Moschandreas D (2006) Indoor environmental quality in schools in relation to academic performance of students: observations of potential contributors to poor IEQ, In: Proceedings of clima 2007, WellBeing Indoors, pp 1–5
6. Gou S, Nik VM, Scartezzini JL, Zhao Q, Li Z (2018) Passive design optimization of newly-built residential buildings in Shanghai for improving indoor thermal comfort while reducing building energy demand. Energy Build 169:484–506
7. Oral GK, Yener AK, Bayazit NT (2004) Building envelope design with the objective to ensure thermal, visual and acoustic comfort conditions. 39(3):281–287. https://doi.org/10.1016/S0360-1323(03)00141-0
8. Yola L, Siong HC, Djaja K (2020) Climatically responsive urban configuration in residential area: Research gaps. AIP Conf Proc 2020(2255):0700141
9. Unver R, Akdag NY, Gedik GZ, Ozturk LD, Karabiber Z (2004) Prediction of building envelope performance in the design stage: an application for office buildings. Build Environ 39:143–152. https://doi.org/10.1016/j.buildenv.2003.08.014

10. Deringer JM, Levine MD, Busch JF (1992) ASEAN-USAID buildings energy conservation project. Final report, volume 3: audits. United States: N. p., Web. https://doi.org/10.2172/10167947
11. Jannat N, Hussien A, Abdullah B, Cotgrave A (2020) A comparative simulation study of the thermal performances of the building envelope wall materials in the tropics. Sustainability 12:4892. https://doi.org/10.3390/su12124892
12. Balaji NC, Mani M, Venkatarama Reddy BV (2013) Thermal performance of the building walls. In: Conference proceeding 1st IBPSA building simulation-2013
13. Vijayalakshmi MM, Natarajan E, Shanmugasundaram V (2006) Thermal behaviour of building wall elements. Int J Appl Sci 6(15):3128–3133 ANSINET publishing
14. Wonorahardjo S, Sutjahja IM, Mardiyati Y, Andoni H, Thomas D, Achsani RA, Steven S (2020) Characterising thermal behaviour of buildings and its effect on urban heat island in tropical areas. Int J Energy Environ Eng 11:129–142
15. Ayegbusi OG (2019) Double skin facade overall thermal transfer value correction factor in hot humid climate. Ph.D. thesis. University Technology Malaysia
16. Si P, Lv Y, Rong X, Shi L, Yan J, Wang X (2020) An innovative building envelope with variable thermal performance for passive heating systems. Appl Energy V 269:0306-2619. https://doi.org/10.1016/j.apenergy.2020.115175
17. US Department of Energy. EnergyPlus Energy Simulation Software. Available at: https://apps1.eere.energy.gov/buildings/energyplus
18. Ben H, Steemers K (2014) Energy retrofit and occupant behaviour in protected housing: a case study of the Brunswick Centre in London. Energy Build 80:120–130
19. Moran F, Blight T, Natarajan S, Shea A (2014) The use of passive house planning package to reduce energy use and CO2 in historic dwellings. Energy Build 75:216–227
20. Jing S et al (2013) Impact of relative humidity on thermal comfort in a warm environment. Indoor Built Environ 22(4):598–607. https://doi.org/10.1177/1420326X12447614
21. ASHRAE (2010) ANSI/ASHRAE Standard 55R-2010—Thermal environmental conditions for human occupancy. Atlanta, American Society of Heating, Refrigerating and Air-Conditioning Engineers, Inc. https://www.ashrae.org/technology/page/548

Identification of Public Open Green Space (RTH) and the Needs for Funding Alternative in DKI Jakarta

Ahmad Fadhil Hidayah, Lin Yola, and Edy Irwansyah

Abstract Identifying the number and area of land based on the type of land parcel ownership for the public green open space (locally known as RTH) facilities and the financing need to meet the realization of targets in DKI Jakarta is the main objective of the research. This study was conducted through the analysis of urban spatial and regulations. The results pinpointed that the planned sub-zone related to public green space was only 14.46%; this figure was lower than the expected target of 20% of the total state area. It was identified that there were 93,748 plots of land and an area of 213,969,971 m^2 for the acquisition of parcels of land for public green space with estimated funding of Rp. 2,261,748,181,458,400. Funding with a combination sourced from the regional expenditure revenue budget (locally known as APBD) and exceeding compensation for building floor coefficients (locally known as KLB) for funding the land acquisition of land parcel for public RTH is certainly not sufficient to meet funding estimates. This study recommended a land consolidation approach and land policy regarding the rearrangement of land tenure and land acquisition efforts for development purposes and environmental protection by promoting community engagement.

Keywords Green open space · Spatial analysis · Zoning regulation

1 Introduction

Green Open Space (RTH) is a part of the urban ecosystem area in the form of open space, where there are many green elements of plants and plants that are intentionally or unintentionally planted, which serve to improve the quality of the city

A. F. Hidayah (✉) · L. Yola
Department of Urban Studies, School of Strategic and Global Studies, Universitas Indonesia, Jakarta, Indonesia
e-mail: ahmad.fadhil91@ui.ac.id

E. Irwansyah
School of Computer Science, Bina Nusantara University, Jakarta, Indonesia

L. Yola et al. (eds.), *Sustainable Architecture and Building Environment*, Lecture Notes in Civil Engineering 161,
https://doi.org/10.1007/978-981-16-2329-5_23

landscape. Garden green space can be utilized by city residents to play, exercise, be creative, socialize, and express media [1].

Urban green space has high life benefits. Various functions related to its existence (ecological, social, economic, and architectural functions) and the aesthetic value it has (objects and environment) can not only improve the quality of the environment and for the survival of urban life but can also be a value of pride and identity of the city. The presence of green space is important in controlling and maintaining environmental integrity and quality. The control of urban development must be carried out proportionally and be in a balance between development and environmental functions. The sustainability of green space in urban areas must be accompanied by the availability and selection of plants following with the direction of the plan and design [2].

The DKI Jakarta Regional Government targets an RTH development plan of 30% which consists of 20% of public green open space and 10% private open green space [3–5]. The target of 20% of public green open space is the target of the Provincial Government of DKI Jakarta involving several Regional Apparatuses, in the form of provision and management of green open space for parks, open green space for tombs and green open space for forests. The area of public open green space covering Park Green Space, Graveyard Green Forest, and RTH Forest until 2017 reached an area of 3076.4 ha, with details for Green Open Green Park covering an area of 2195.3 ha, Grave Green Open Area covering 626.83 ha and Forest Green Opening area of 254.26 ha. In total, the number of public green open space in Jakarta is only 9.98% of the expected 20% of public green open space targets. While the Private Green Open Space has reached the target of 10% [6].

The purpose of this study is to identify the number and extent of land based on the type of land ownership parcel of land for public green space (RTH) and financing to meet the target realization in DKI Jakarta Province.

2 Methods

The research method consists of three stages, namely the study of literature, spatial and non-spatial data collection, and data analysis. A literature study is carried out with a literature search containing reviews and summaries which include journals, reports, laws and regulations, and other information documents related to Land Identification for public green space facilities and alternative funding in DKI Jakarta Province. Spatial data collected consists of operational maps of DKI Jakarta RDTR and PZ as well as spatial data on land parcel in Jakarta in 2016–2017 obtained from the National Land Agency (BPN). Both types of data are collected in the shapefile (shp) format. Data analysis consists of spatial analysis to calculate the need for green space and analysis of alternative funding for space requirements for green space provision. Spatial analysis conducted using a geographical information system (GIS) tools include a query and spatial join operation. A query is to display data from a database by calling from the tables in the database, with not all tables

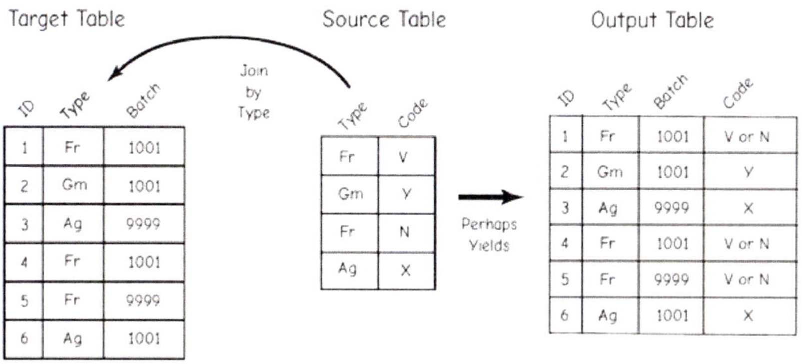

Fig. 1 Simple selection and selection operator in GIS and the example of joint feature operation in GIS [7]

being displayed. In Fig. 1 we can see examples of spatial queries that are implemented with simple selection and AND selection operators in GIS.

In Fig. 1 that are displayed are only those that correspond to the data table that we want. Spatial join is the process of combining tabular data with a join function. This process combines tabular target feature/layer data to which data will be added with a join feature which is a feature/table that will be added. This process will produce new tabular data which is the result of the combination of 2 (two) tables using the merging process option.

3 Result and Discussion

3.1 Public RTH Based on RDTR PZ DKI Jakarta Province (Development Right) Land Ownership of Land Parcel in DKI Jakarta (Property Right)

Classification of zoning related to public RTH in RDTR and PZ includes: (1) a protection zone consisting of asylum and nature conservation (L.1) sub-zones, protected border zones (L.2), and island core conservation sub-zones (L.3); (2) the urban forest zone consists of the urban forest sub-zone (H.1); (3) the urban/environmental park zone consists of the urban/environmental park sub zone (H.2); (4) the burial zone consists of the burial sub-zone (H.3); (5) the green lane zone consists of the green lane sub-zone (H.4), the high-voltage green sub-zone (H.5), and the railroad green safety zone (H.6); (6) the recreation green zone consists of the recreation green sub-zone (H.7); and (7) green open zones of cultivation in the island region which consist of green open zones of island cultivation (H.8). The spatial and broad distribution, as well as the percentage of each public green space sub-zone as can be seen in Fig. 2 and Table 1.

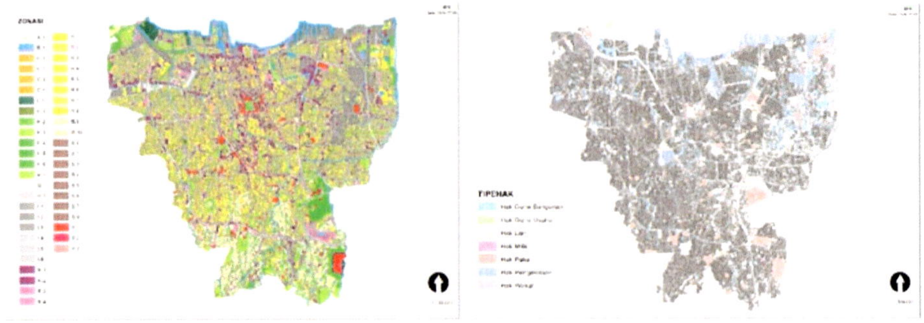

Fig. 2 Zoning map of DKI Jakarta and location distribution map of land parcel in DKI Jakarta

Table 1 Area and percentage of DKI Jakarta provincial sub-zone

No	Sub zona	Area and percentage of Sub zones	
		Area (in ha)	Percentage
1	H.1	17.03	0.03
2	H.2	1566.64	2.97
3	H.3	585.34	1.11
4	H.4	3729.69	7.07
5	H.5	428.66	0.81
6	H.6	295.03	0.56
7	H.7	640.34	1.21
8	L.1	369.17	0.7

As of 2016, there were 1,028,736 parcels of land spread across 224 urban villages throughout the DKI Jakarta Province (Fig. 2).

Based on the results of the calculation of the area for each sub-zone in the public green space area which includes the asylum and nature conservation sub-zone (L.1), urban forest sub-zone (H.1), urban/environmental park sub-zone (H.2), burial sub-zone (H.3), green line sub-zone (H.4), high voltage green sub-zone (H.5), and railroad safety sub-zone (H.6) and the recreation green sub-zone (H.7), the total area is only around 14.46% of the total area of the city and has not yet reached the 20% target set for RTH public.

3.2 Comparison of Development Right and Property Right

Comparison of the area of development rights and property right in DKI, Jakarta is done to get a picture of the appropriateness of the designation, especially for public green space. Illustration in the Kapuk Cengkareng in the northern part of the Jakarta area, most of the public RTH plan has not yet to be realized because there are still

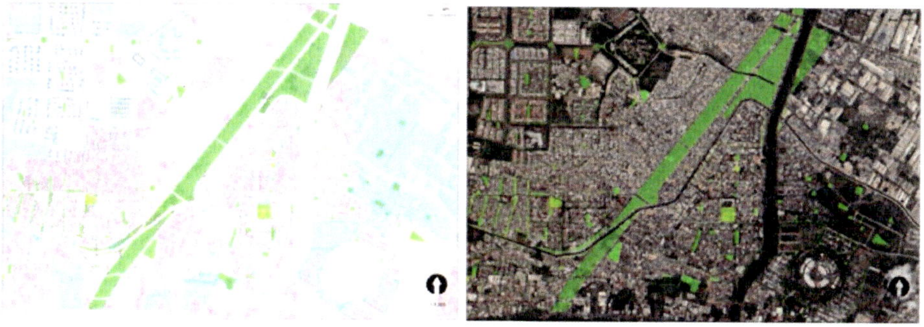

Fig. 3 Public RTH and type of ownership of land parcel (Kapuk Area, Northern Part of Jakarta)

many dwellings in the same location (Fig. 3). Illustration show that there is still a lot of land parcels that do not yet have a type of land parcels certificates that are in the public RTH zoning. The condition of the public RTH plan that has not been realized and in the existing conditions in the form of the residential population is also widely spread throughout the DKI Jakarta Province.

Analysis in all regions of DKI Jakarta can be obtained from the fact that East Jakarta is an administrative area which has the most number of land parcels in the public RTH, which is 33,739 land parcels. North Jakarta is the administrative area in DKI Jakarta with the largest number of land parcels located in the public RTH, covering an area of 695,568,281 m^2. Based on the type of ownership rights, other rights and building use rights identified are the highest in the public green space zone, each covering an area of 67,094 m^2; 16,246 m^2; and 10,408 m^2 land parcels. Specifically for the type of land parcels, management rights were identified as having the highest number of parcels in the public RTH zone, covering 682,515,184 m^2.

3.3 Estimated Funding for Land Parcel Acquisition in Public RTH

The provision of land parcels will be made on the three broadest types of rights that are in the public RTH zone, namely building use rights, ownership rights, and other rights. Specifically for use rights and management rights, land acquisition parcels cannot be acquired for public green space, because in principle it belongs to the state.

When referring to the sample of the distribution of the value of tax objects (NJOP) of DKI Jakarta Province in 2018, assuming an average amount of Rp. 10,570,400 [8], funding is needed for land acquisition parcels of land for public RTH on the type of building rights, ownership rights, and other rights amounting to Rp. 2,261,748,181,458,400.

3.4 Funding Alternative Land Acquisition of Land Parcel for Public RTH

Alternative funding for land parcel acquisition for public RTH can be done with two events, namely by using the budget and revenue, regional expenditure (APBD) or through compensation for exceeding building floor coefficients (KLB).

In 2017, 2018, and 2019 DKI Jakarta Province APBDs respectively amounted to Rp. 71,832,575,363,258, Rp. 77,117,365,231,898 and Rp. 89,088,351,842,504 [9]. This amount is certain to be insufficient for funding land parcel acquisition for public RTH based on the estimated calculation of Rp. 2,261,748,181,458,400.

High KLB values can be given at locations designated as bonus zoning regulation (TPZ) technical zones (code a) with zone delineation as stipulated in regional regulation No. 1/2014 concerning RDTR and PZ [10–12], compensation is determined by providing public facilities such as (a) providing land and/or building public RTH; (b) provide land and/or build public flats; (c) provide land and/or build reservoirs or sites; (d) providing infrastructure (e) provide pathways and improve the quality of pedestrian facilities that are integrated with public transportation; and (f) providing bicycle lanes integrated with public transportation.

The form of compensation for the acquisition of parcels of public RTH is required to be transferred to the Provincial Government of DKI Jakarta to be recorded as an asset. For the form of compensation to be submitted, valuation is conducted first by the Public Appraisal Service Office (KJPP). Like the APBD, land acquisition parcels of land for public RTH from compensation for the excess of KLB has also been confirmed to be inadequate based on estimated calculations (Rp. 2,261,748,181,458,400).

4 Conclusions

This study investigated DKI Jakarta's feasibility to implement the 20% target of public open green space by using the spatial and regulation analysis. This study reported that the sub-zones related to public RTH in the plan were 14.46% over the target of 20%. This study presented the funding need analysis for the acquisition of land parcels in the amount of Rp. 2,261,748,181,458,400 for land parcel acquisition for public RTH by referring to the NJOP Distribution sample for DKI Jakarta Province in 2018. A mixed source of funding from the APBD and KLB to finance land parcels' acquisition/provision for public RTH was reported insufficient to meet the estimated funding.

This study's finding is vital literature for the policymakers and planners to fulfill the current plan in creating a better environment and greener and sustainable DKI Jakarta.

References

1. Prakoso P, Herdiansyah H (2019) Analisis Implementasi 30% Ruang Terbuka Hijau Di DKI Jakarta: Majalah Ilmiah Globe 21(1):17–26
2. Wibowo A, Ritonga M (2016) Kebutuhan Pengembangan Standar Nasional Indonesia Fasilitas Taman Kota. J Standardisasi 18(3):161–170
3. Government of Indonesia (2007) Undang-Undang Republik Indonesia Nomor 26 Tahun 2007 Tentang Penataan Ruang. Lembaran Negara Republik Indonesia Tahun 2007 Nomor 68. Sekretariat Negara Republik Indonesia. Jakarta
4. Government of Indonesia (2014) Peraturan Menteri Pekerjaan Umum No. 01/PRT/M/2014 Tahun 2014 Tentang Standar Pelayanan Minimal Bidang Pekerjaan Umum dan Penataan Ruang. Lembaran Negara Republik Indonesia Tahun 2014 Nomor 267. Sekretariat Negara Republik Indonesia. Jakarta
5. Provincial Government of DKI Jakarta (2012) Peraturan Daerah Provinsi DKI Jakarta No. 1 Tahun 2012 Tentang Rencana Tata Ruang Wilayah 2030. Lembaran Daerah Provinsi DKI Jakarta Tahun 2012 Nomor 1. Sekretariat Daerah Provinsi DKI Jakarta. Jakarta
6. Provincial Government of DKI Jakarta (2018) Rencana Pembangunan Jangka Menengah Daerah (RPJMD) Provinsi DKI Jakarta Tahun 2017-2022. Jakarta
7. Bolstad, Paul (2016) GIS fundamental: a first text on geographic information systems. XanEdu. ISBN: 978-1-50669-587-7
8. Provincial Government of DKI Jakarta (2018) Peraturan Gubernur Provinsi DKI Jakarta Nomor 24 Tahun 2018 Tentang Penetapan Nilai Jual Objek Pajak Bumi Dan Bangunan Perdesaan Dan Perkotaan Tahun 2018. Sekretariat Daerah Provinsi DKI Jakarta. Jakarta
9. Provincial Government of DKI Jakarta (2019) Anggaran Pendapatan dan Belanja Daerah Provinsi DKI Jakarta. https://apbd.jakarta.go.id. Accessed 6 Apr 2019
10. Provincial Government of DKI Jakarta (2014) Peraturan Daerah Provinsi DKI Jakarta Nomor 1 Tahun 2014 Tentang Rencana Detail Tata Ruang dan Peraturan Zonasi. Lembaran Daerah Provinsi DKI Jakarta Tahun 2014 Nomor 301. Sekretariat Daerah Provinsi DKI Jakarta. Jakarta
11. Provincial Government of DKI Jakarta (2016) Peraturan Gubernur Provinsi DKI Jakarta Nomor 210 Tahun 2016 Tentang Pengenaan Kompensasi Terhadap Pelampauan Nilai Koefisien Lantai Bangunan. Berita Daerah Provinsi DKI Jakarta Tahun 2016 Nomor 63010. Sekretariat Daerah Provinsi DKI Jakarta. Jakarta
12. Zulkaidi D (2018) Teknik-Teknik Pengaturan Zonasi dan Pertimbangan Penerapannya. Bahan Pendidikan Pusat Penelitian Infrastruktur dan Kewilayahan. Institut Teknologi Bandung dan Ikatan Ahli Perencanaan Provinsi Jawa Barat. Bandung, 26–28 Sept 2018